GLOBAL
WEATHER
PREDICTION

The Coming Revolution

GLOBAL WEATHER PREDICTION

The Coming Revolution

EDITED BY
BRUCE LUSIGNAN
AND
JOHN KIELY
Stanford University

HOLT, RINEHART and WINSTON, Inc.

New York   Chicago   San Francisco   Atlanta
Dallas   Montreal   Toronto   London   Sydney

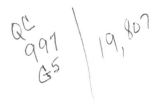

Preface

WITHIN the next decade, there will be a revolution in the science of weather prediction, for within five to ten years it should be possible to make accurate 10- to 14-day weather forecasts. This amounts to a quantum jump beyond the current forecasting ability, and its benefits will involve saving millions of dollars and hundreds of lives which are lost each year in unexpected floods and storms, and will have far-reaching consequences for large segments of the world's population. Hence, a book about this revolution, written by some of the men who make it possible, may be of interest to a wide range of readers.

This book grew out of a series of some 30 lectures given over a period of four months in 1966 to a graduate Space Systems Engineering course at Stanford University. The goal of the class, which consisted of some 50 engineers, lawyers, businessmen, economists, and one philosopher, was to design a global weather satellite system. Since a knowledge of meteorology was not a prerequisite for the course, the aim of the lectures was to provide an extensive introduction into the many aspects of a global weather prediction system. The members of the class then proceeded with their own research. The final weather satellite system design has been published by Stanford in a volume entitled *SPINMAP*, sections of which are included in this book.

Among the lecturers were many of the leading meteorologists, physicists, businessmen, and statesmen involved in the field of weather prediction. Among the many topics they dealt with were the history and future of meteorology, the costs and benefits of improved weather forecasts, the TIROS and Nimbus satellite systems, numerical models for the atmosphere, and the use of the laser for atmospheric observations. This book came about as a result of the excitement generated by the individual concepts and by the overall possibilities for the next decade which grew out of these lectures.

Many people with a diversity of backgrounds enjoyed these lectures, and they have been edited into the articles of this book for readers who are interested in the concepts of weather prediction in the 1970's. The major portion of the book has been written for readers with a general engineering or scientific background. However, a substantial part of the first five parts should be intelligible to any man of letters, while parts of Part 6, on numerical models, require a reasonable mathematical competency.

The articles of the book fall naturally into two unequal sections. The first, consisting of Parts 1 to 3, provides a general introduction to the concepts, problems, cost, and benefits of weather prediction, while the second, Parts 4 to 7, describes the technology, the satellite data systems, and the models which will enable man to comprehend the motions and properties of the atmosphere on a global basis well enough to begin to make the long-term forecasts.

The problem of measuring such simple variables as temperature, pressure, and water vapor content at a certain point in the atmosphere remotely from an orbiting satellite has yet to be completely solved. Part 7 contains a discussion of problems of this nature together with a description of the French EOLE weather satellite system by Professor Pierre Morel of the University of Paris.

The 18 articles in this book cover a wide range of topics and represent the thinking and research of the individual authors. This book is not a text in meteorology, nor is it a comprehensive design of a weather forecasting system, but rather its aim is to present an overall view of the exciting possibilities as well as the technical and political problems involved in bringing about the revolution in weather prediction.

The editors wish to thank the 16 authors who gave the initial lectures and who gave freely of their time and interest to the sometimes difficult job of transforming the spoken word of a lecture to the written word of an article for a book. We are also grateful to the Department of Engineering of Stanford University for their generous support of the Space Systems Engineering course within which this book had its beginnings and grew to maturity. We also wish to thank Wendy Meara, Jan Jeffery, Betty Griffiths, Mary Beard, and Sally Burns for their advice and interest in the typing and preparation of the manuscript.

Bruce Lusignan
John Kiely

Contents

Acronyms, Titles,
and Unit Abbreviations

APT
    Automatic Picture Transmission
Camera systems on board TIROS and other meteorological satellites

ATS
    Applications Technology Satellites
A series of multipurpose research satellites

AVCS
    Advanced Vidicon Camera System
Used first on TOS

BUV
    Backscatter Ultraviolet Spectrometer

CDA
    Command and Data Acquisition Station
(TIROS)

COMSAT
    An early Communications Satellite Project which developed into SYNCOM (Part 4, Rosen)

COSPAR
    Committee on Space Research

CRT
    Cathode Ray Tube

dB
    Decibel

Early Bird
    A spin stabilized synchronous communications satellite in an equitorial orbit

ekW
    Electrical kilowatt

EOLE
    A French weather data system using free floating constant level balloons and a tracking satellite (similar to the U.S. GHOST system)

ESSA
    Environmental Science Services Administration. Washington, D.C.

ETS	Educational Television Satellite
eV	Electronvolt
GARP	Global Atmospheric Research Program
GHOST	Global Horizontal Sounding Technique *A balloon satellite data system sponsored by the U.S. Weather Bureau*
TROMEX GLOMEX	Global Meteorological Experiments
HRIR	High Resolution Infrared Radiometer
Hz	Hertz (cycles per second)
ICAU	A treaty for International Commercial Aircraft Operations
ICSU	International Council of Scientific Unions
IR	Infrared *Radiation with wavelengths just longer than visible light*
km	Kilometer
kW	Kilowatt
LARC	*A computer used at the Lawrence Radiation Laboratory (Part 6, Leith)*
LIDAR	Laser Radar *An instrument capable of measuring the radiation backscattered from a ruby laser beam (Part 7, Ligda)*
MHz	Megahertz
mb	Millibar
MRIR	Medium Resolution Infrared Radiometer
NASA	National Aeronautics and Space Administration
NCAR	The National Center for Atmospheric Research
NESC	National Environment Satellite Center in Suitland, Maryland

Nimbus	*A meteorological observatory spacecraft (a successor to the TIROS/TOS series)*
NOMAD	*A Navy Buoy anchored in the Gulf of Mexico used for gathering and transmitting meteorological data*
nmi	Nautical miles
OAO	Orbiting Astronomical Observatory
OMEGA	*A U.S. Navy navigational service*
RTG	Radioisotope Thermoelectric Generator (*SNAP-19*)
SIRS	Satellite Infrared Spectrometer
SNAP	*A series of nuclear thermoelectric generators for satellite power supplies*
Solomon STRETCH	*Computers used with the numerical models of the atmosphere*
SPINMAP	Stanford Proposal for an International Network for Meteorological Analysis and Prediction
SSCC	Spin-Scan Cloud Camera
SYNCOM	A Spin Stabilized Synchronous Communication Satellite
TIREC	TIROS Ice Reconaissance Program
TIROS	*An early weather satellite system*
TOS	TIROS Operational System
WMO	World Meteorological Organization (*An agency of the United Nations*)
WWW	The World Weather Watch

PART 1

introduction

THE weather affects all aspects of man's life. It determines when farmers plant and harvest their crops; it affects the price a housewife pays for fruits and vegetables; it determines what route an ocean liner or an airplane should take between cities; it influences the timing of a manned space shot from Cape Kennedy; it determines whether we take an umbrella or an overcoat to work. Fuel companies must decide whether a coming winter will be harsh or mild to estimate the size of their coal and fuel-oil reserves. Construction companies, picnickers, field generals, farmers cutting hay, and brides-to-be would all very much like to know whether it will rain next Tuesday or two weeks from today.

Hence, for thousands of years man has been trying with every means at his disposal to predict, change, and cajole the moving masses of air and moisture which make up the atmosphere. But for all these years the movement of storms and the large-scale energy sources and sinks of the atmosphere have remained a mystery.

Today, however, we are on the threshold of a new era when global satellite and balloon data systems together with sophisticated atmospheric models and high-speed computers will enable man to begin to understand the complex motions and energies of the atmosphere and hence to begin to make accurate long-range weather predictions. Within a decade it should be possible to make 10-to-14-day weather forecasts with an accuracy greater than that of the 2-to-3-day forecasts we now receive. The following articles attempt to describe various facets of this new era.

The word "meteorology" will come up frequently in this book. Meteorology as we use the term is concerned with the physics, chemistry, and dynamics of the atmosphere and with its effects on the oceans, the earth's surface, and life in general. It assumes a physical understanding of the atmosphere, and it strives toward accurate prediction and eventual control of the atmospheric environment.

In this book meteorology is viewed primarily from the standpoint of

3

weather prediction and its supporting research programs. Bounded by the ocean and solid earth below and by the ionosphere above, the "meteorological atmosphere" considered here extends to about 50 miles above the earth's surface.

One of the most troublesome characteristics of storms and high-and low-pressure areas is that they move over large areas in relatively short periods of time. A certain mass of air may travel around the globe two, three, or even four times in a period of two weeks. Now for several centuries men have been able to measure the basic meteorological variables of pressure, temperature, and water-vapor content at the surface of the earth. Such measurements have of necessity been taken in the scattered centers of population. There have been very few measurements taken in uninhabited areas and over the three quarters of the earth covered by oceans, and until quite recently none in the important upper levels of the atmosphere. It has been estimated that less than 20 percent of the earth's surface is adequately covered by upper-air observing stations, and unless this situation is corrected, it is hopeless to expect a substantial advancement in weather forecasting. Meteorologists have been making weather forecasts with data which are marginally acceptable for short-range forecasts but which would be seriously inadequate for longer-range predictions.

In addition to this lack of data meteorologists have had only an imperfect understanding of the physics of the atmosphere. The atmosphere is basically an initial state problem; that is, if one knew the initial temperature, pressure, and velocities of the air over all the earth, then by solving a certain set of differential equations it should be possible to predict the entire future pattern of air movement. To a certain extent this can now be done with the atmospheric models available (Part 6). However, there are an enormous number of complexities (such as the turbulent energy exchange between large- and small-scale weather systems) which quickly make the predictions of the model invalid.

The goal of the next decade is to remedy these two obstacles; to set up a global data-collection system and to perfect our physical understanding of the atmosphere. There are numerous other problems, such as communication links, the optimum form for presenting the data, and rapid preparation and dissemination of local forecasts to individual users, but these require wise administrative decisions rather than the more lengthy development of new techniques.

It is generally accepted that for weather predictions in excess of a few days, the earth's atmosphere must be treated as a single dynamic system. At present, numerical forecasts are prepared routinely for periods of three to four days, and for areas covering about one-third of the earth. To extend forecasts to longer periods or to larger areas requires knowledge of the initial state of the atmosphere on a global or at least hemispheric scale. Otherwise, unknown disturbances will migrate into the prediction areas and contaminate the forecast.

Ultimately it is likely that the global meteorological network will consist of a system of satellites to gather data throughout the atmosphere remotely, tracking stations, and one or more large meteorological centers which will use the atmospheric models and the largest computers available to analyze the satellite data and to prepare the predictions. Local weather forecasts for periods ranging from 6 hours to 14 days could then be sent to the various national or regional distribution centers.

For the immediate future, however, it does not appear feasible to measure temperature, pressure, wind velocity, or water-vapor content remotely from satellites. The various technologies have been proposed and widely discussed, but it will be several years before such systems are operable. All the systems now being built involve some intermediate network present in the atmosphere to collect data and transmit it either directly to earth or to monitoring satellites. The French EOLE system and the United States GHOST systems both propose the use of light-weight free-floating constant-level balloons. Part 7 contains descriptions of these systems by Pierre Morel and Stanley Ruttenberg.

In all the numerical models now in use the atmosphere is divided up into boxes and data is read in at each of the corners or grid points. The model uses a numerical integration scheme to advance the variables at each grid point through one time step. This basic process is repeated over and over again to predict a future state of the atmosphere. The choice of the box or mesh size and the time step are two of the most important parameters of the entire network.

The mesh size determines the size of atmospheric phenomena which may be dealt with in the model. For instance, with a box of 1000 km on a side one could not hope to predict the occurrence of any storms whose dimensions were less than 1000 km because the model receives data at only one point in every box—every 1000 km. Hence the weather maps produced by the model might predict a large low-pressure area or a hurricane but they would never be able to predict a tornado or a thunderstorm.

From this viewpoint it is desirable to have the smallest possible grid spacing. Ideally this might be only a few kilometers. However, the amount of data and the amount of computer time required to advance the variables through one time step increase rapidly with the number of grid points.

At present, for a global model, a grid spacing of 300 to 500 km is at the upper limit of computer capabilities. This figure will of course become smaller as larger and faster machines are available. The time step for each integration affects, among other things, the accuracy of the predicted states. At present, for the computer calculation to run at about 10 times faster than real time (which is a lower limit if the model is to make useable predictions) a time step of about 10 minutes is generally used.

Most of the work now going on is devoted to developing hardware, building systems, and testing systems. The pioneer TIROS and Nimbus satellite systems, in addition to sending back a wealth of cloud-cover pictures, have been testing many of the cameras and sensors which will ultimately be used on the data satellites. The development of the constant-level balloons and the ultra-lightweight electronic circuitry (to eliminate the danger of collisions with high flying aircraft) is essentially complete. However, as mentioned earlier, there is a great deal of work yet to be done on the remote sensors to measure temperature, pressure, and wind velocities from orbiting satellites.

The global weather-prediction network is still in its infancy, but it is a healthy child and growing rapidly. The articles which follow describe its current appearance and indicate in some detail the directions of its future development.

PART 2

goals

THOMAS MALONE

Thomas Malone is Vice-President and Director of Research for the Travelers Insurance Company, Hartford, Connecticut. He is chairman of the Committee on Atmospheric Sciences of the National Academy of Science and served as President of the American Meteorological Society from 1960 to 1962 and as President of the American Geophysical Union from 1961 to 1964.

Abstract—Dr. Malone begins the book with "An Overview" of the history, current developments, and future requirements of a global weather system. He ends with a suggestion, in a quote from the late President Kennedy, that the international effort now underway in meteorology may have implications beyond the boundaries of the sciences involved: "a manageable, worthwhile system of World order will be based 'not on a sudden revolution of human nature, but on a gradual evolution in human institutions'."

An Overview

IN a memorable address before the United Nations in September 1961, President John F. Kennedy proposed a four-point program for the peaceful use of outer space, to be developed under the auspices of the United Nations. One part of the program consisted of an international collaborative effort "in weather prediction and eventually in weather control." The General Assembly responded with a unanimous resolution calling upon member states and governmental and nongovernmental organizations to develop programs which would advance the state of the atmospheric sciences and which would lead to a surer knowledge of the basic physical forces affecting climate and to the improvement of weather forecasting and the exploration of the possibilities and limitations of large-scale weather modification. These resolutions stimulated extensive discussion, study, and planning within governmental and nongovernmental circles in this country and abroad.

From these discussions, the twin concepts of a World Weather Watch (WWW) and a Global Atmospheric Research Program (GARP) have emerged. The objectives of WWW are to bring the global atmosphere under observational surveillance, to establish a worldwide telecommunication system for transmitting and collecting weather observations, and to bring into being a set of World Meteorological Centers and Regional Meteorological Centers equipped with computational facilities of sufficient speed and capacity to process the global meteorological data in a manner that will serve the needs of weather prediction, research and climatological archiving. The objective of GARP is to develop an understanding of the global atmospheric circulation in sufficient depth to permit the mathematical/physical prediction of weather two or more weeks in advance, to provide the basis for a physical explanation of world climate, and to make possible the scientific exploration of the possibilities and limitations of large-scale climate modification.

A measure of the technological challenge implicit in this international effort is provided by the estimate that only 20 percent of the global atmosphere is being adequately observed at present by the mosaic of 130 national meteorological observation systems. A complex array of earth-orbiting satellites, ocean buoys, unattended remote land stations and freely floating balloons—all linked together by a high-speed communication system—will be required to close the gap between what is presently observed and that which must be observed to achieve the objectives of the program. To process in timely fashion the volume of data that will be generated will require computers 100 times faster than those currently available.

This two-pronged international program will proceed in four interrelated phases.

The first phase will involve the application of proved technology in observations and communications to the augmentation of existing national systems. World meteorological centers are being established in the United States, the Soviet Union, and Australia to facilitate the gathering and exchange of meteorological data and to utilize current computers in preparing weather analyses and forecasts.

The second phase will proceed simultaneously with the first and will involve research and development on the new technology envisioned for operational implementation in the early 1970's. This new technology will include new generations of meteorological satellites, automatic meteorological ocean buoys, communications satellites, horizontal sounding balloons, a new generation of electronic computers, and more sophisticated techniques for processing meteorological data.

The third phase is concerned with research on the global wind systems that are identified in the language of the meteorologist as "the general circulation." This research includes identification of the specific components of the problem of the general circulation, determination of the data requirements for research purposes, organization of a series of regional field experiments and, finally, a sustained effort extending over about a year during which special observations will be required to conduct a full-scale research program on the general circulation. From this endeavor, it is anticipated that significant improvements will be made in weather-prediction techniques and substantial increases will be achieved in the span of time for which physically based predictions are valid.

The fourth phase will represent the translation of the research results into observational and prediction techniques employed operationally in the World Weather Watch. Implementation of this phase will most likely come about by a process of evolution rather than in a discrete step. The full impact of the program should begin to be felt during the latter part of the 1970's.

To view these programs in the proper perspective, it is helpful to note that progress toward an understanding of our atmospheric environment has taken place somewhat unevenly over the past 2500 years. Three significant turning points can be identified. A breakthrough of a sort was achieved by the perspicacious Greeks, who lifted study of the atmosphere out of the context of mythology, in which weather was ascribed to the whims of a host of deities, and placed it on a level characterized by a scientific attitude of reason and a physical interpretation of systematic observations. About 350 B.C., Aristotle was able to summarize all existing knowledge in his *Meteorologica,* a compendium that remained the authoritative reference work in this field for nearly 2000 years.

A second significant advance occurred with the invention of the thermometer, the barometer, and the hygrometer during the 16th and 17th centuries. Extension of the operational power of the human eye by instrumentation which can measure parameters the eye cannot discern opened the first quantitative era in meteorology: a period when discrete measurements were analyzed carefully to gain a better understanding of the laws of nature. The researches of many great scientists were concerned directly or indirectly with atmospheric problems. Galileo, Evangelista Toricelli, Blaise Pascal, Robert Boyle, Jacques A. César Charles, René Descartes, Isaac Newton, Anders Celsius, Benjamin Franklin, Joseph Black, Joseph Louis LaGrange, Pierre Simon de Laplace, John Dalton, and many others left their marks in meteorology as they became intrigued with the explanation of physical phenomena in the atmosphere.

The third major development took place during the 19th century, when Brandes, in 1820, prepared the first weather map in an attempt to portray an integrated picture of the atmosphere on a synoptic basis. The effort was premature, but it did provide a basis for the preoccupation during the latter half of the 19th century with the task of depicting successive states of the atmosphere with the objective of preparing weather forecasts. In one sense, the shift in emphasis from *why* the weather to *what will be* the weather was an unfortunate diversification of effort with the result that scientific progress was rather slow during the first part of the 20th century. In a broader sense, however, this was a necessary and even desirable step in the scientific advance of meteorology because the economic importance of even an unsatisfactory weather forecast made possible the national and international networks of observing stations required to view the atmosphere in its global dimension. From these data and from fragmentary and scattered studies on the fundamental physics of atmospheric processes, there has begun to emerge during the past two or three decades a reasonably coherent picture of the scientific problem in learning about the lower portion of the atmosphere. The fact that elements of the problem can even be formulated is perhaps the most exciting development in meteorology over the past 3000 years.

With respect to the scientific problem, we may think of the atmosphere as a complex physical system in which movement of air, changes in temperature, and transformation of water among the liquid, solid, and gaseous phases are of considerable practical interest, all taking place in response to certain forces or through particular processes. Although the atmosphere is far from being a tidy little deterministic system, in principle, we can cast these processes in quantitative form and relate a given state of the atmosphere to a subsequent state if there is no human intervention in these processes. This is a basis for weather prediction. Human intervention constitutes weather modification. In this sense, the matters of weather prediction and weather modification are meaningful scientific problems.

They are, however, complicated problems. The earth's atmosphere may be viewed as an envelope rotating with the earth as well as relative to it. The relative motion is caused by the forces associated with the rotation of the earth, and forces associated with the sources and sinks of energy that are variable in number, location, and strength. These sources and sinks of energy depend on the distribution of shortwave solar radiation, the flux of outgoing long-wave radiation, the latent heat involved in the change in phase of water, the transfer of sensible heat between the atmosphere and the underlying surface, and finally the air motion itself. The kinetic energy of air motion exists in an array of scale sizes that extend from planetary wave systems down to molecular movement. There is continuous exchange of kinetic energy from one scale to another, and the kinetic energy is continually being exchanged with other forms of energy in the atmosphere.

A characteristic of the atmosphere that frustrates the weather forecaster but provides a basis for optimism for the weather modifier is the tendency for the processes in the atmosphere to demonstrate certain traits of instability. This tendency is readily apparent, from everyday experience, in the tendency for the amplitude of atmospheric disturbances to increase with time. For example, a small puffy cloud may grow to a towering thunderstorm in a matter of hours; a gentle zephyr in tropical latitudes may develop into a "killer" hurricane in a matter of days; and a small low-pressure center may grow to a vigorous extratropical cyclone within a single day.

We are just beginning to understand (1) the instability of supercooled water droplets which, when released, provide a local source of sensible energy, (2) the convective instability of a rising current of air within which water vapor is condensing into liquid, thus affecting the vertical distribution of sensible energy, and (3) the so-called baroclinic instability of the large-scale, planetary atmospheric waves, which when released can profoundly alter the nature of the great global system of winds.

Atmospheric instability is one of the principal reasons why weather predictions have come to be couched in probabilistic terms. Since this instability almost invariably leads to some uncertainty in the forecast, the statement that there is a 70 percent chance of rainfall is more useful in weather prediction for the purposes of operational decisions than a simple forecast that rain is expected. Similarly, the possibility that large effects may be produced from relatively modest but highly selective human interventions opens up the possibility that weather and climate modification may some day be operationally feasible.

Within the context of this general background, it is pertinent to mention three scientific and technological developments of recent years that have opened up new dimensions of scientific research and given special relevance to the exploration of new patterns of international cooperation:

First, understanding of the physical processes occurring in the atmosphere has now progressed to the point at which they can be expressed in equations that constitute mathematical models. These models permit simulation of natural processes useful both in the perfection of prediction techniques and in the assessment of the consequences of human intervention in these natural processes. Although crude and oversimplified relative to the processes they are intended to simulate, useful models have been constructed of atmospheric phenomena that range in size from a single cloud to circulation of air over an entire hemisphere. There is almost unlimited potential for extension and refinement. Several of these models are discussed in Part 6.

Second, the advent of high-speed electronic computers has hastened the possibility of integrating the nonlinear, partial, differential equations governing atmospheric motions by numerical methods. Computers, in turn, have provided a powerful new tool for the growing number of investigators who are seeking to understand atmospheric processes by means of analysis of the relevant mathematical equations and who are anxious to perfect these methods for numerical weather prediction.

Third, expanding the capabilities of making the observations and measurements specifying the initial and final atmospheric conditions that must be reconciled by the computerized atmospheric models if they are to be meaningful. These emerging capabilities, which are discussed in Part 5, range from the dramatic demonstration of the utility of meteorologic satellites on a global scale to the precision power of intricate measurements of the relevant physical characteristics in a single cloud.

Taken together, these three advances have set the stage for the development of quantitative techniques of prediction firmly founded on a physical basis. Moreover, they mark the transition in the problem of weather modification from an era of intellectually undisciplined speculation and more or less opportunistic field experimentation into an era of rational, organized inquiry in which a set of meaningful scientific questions can be explored analytically and by means of carefully designed field experiments. Quite clearly, within the next decade or so it will become possible to explore, through simulation techniques, an almost unlimited array of deliberate interventions in natural atmospheric processes, and to assess possibilities and limitations. These studies will inevitably lead to specific requirements for meteorological measurements that will deepen our understanding of natural processes. As an example, mathematical models of the atmosphere have already been used in a preliminary way to assess the consequences of the inadvertent intervention associated with the increase of atmospheric carbon dioxide. Models may yet be used to define the tolerable limits to this large-scale geophysical experiment that mankind is undertaking, or, alternatively, to determine desirable countervailing measures.

It is against this background that the articles in this book should be read and understood.

Progress toward implementation of the concepts behind WWW and GARP is accelerating. The year 1967 was a particularly auspicious one for these programs. The World Meteorological Organization (a specialized agency of the United Nations) held its fifth congress in Geneva in April 1967 and approved a detailed set of resolutions which spell out the essential elements of the World Weather Watch. The significance of this step lay in the fact that the governments of the 130 member nations had made a firm commitment to the program.

Meanwhile, activity in the nongovernmental international organizations had been building up to a high pitch. A special Committee on Atmospheric Sciences was established by the International Council of Scientific Unions (ICSU) under the aegis of the International Union of Geodesy and Geophysics, and this committee worked closely with a specially designated Working Group VI which was established by ICSU under the aegis of the Committee on Space Research (COSPAR). A series of meetings between these two groups and the Advisory Committee of WMO during 1965 and 1966 culminated in a study conference held in Stockholm during the summer of 1967. This conference was attended by more than 50 scientists from 13 interested nations. The objective of the conference was to develop the general design for GARP. It brought together specialists in the large-scale dynamics of the atmosphere; experts familiar with the boundary-layer fluxes and the problems of air-sea interaction; specialists in the dynamical problems of the tropics, convection processes, and mesoscale phenomenon; and scientists whose research was on problems of atmospheric radiation with a group of scientists and engineers competent in the field of instrumentation and space technology. The result of the conference was a 144-page report which represented a consensus of the conference participants.

This research program was presented to the International Association of Meteorology and Atmospheric Physics at its general assembly held in Lucerne in the fall of 1967. Not only was the research program approved, but a resolution was also adopted establishing on behalf of WMO and ICSU a Joint Organizing Committee with the following functions:

> To consider, to endorse and to recommend jointly to ICSU and WMO scientific goals and plans for GARP, including GARP Sub-Programmes, that are considered essential prerequisites in defining the scientific requirements of GARP (this will include defining detailed experimental objectives and operational requirements for their implementation).
> To recommend to WMO those techniques and procedures developed in GARP programmes that may be applied in the operation of WWW.
> To recommend to WMO the manner in which the scientific requirements of GARP can best be supported by the operation of WWW.

These actions were acted upon favorably by the Council of the International Union of Geodesy and Geophysics and by the Executive Committee of ICSU. The Joint Organizing Committee consisting of 12 distinguished scientists from all over the world was appointed in late 1967 with plans to hold an organizing meeting early in 1968.

The United States Committee for GARP, under the chairmanship of the distinguished scientist Professor Jule Charney of MIT was established by the National Academy of Sciences in early 1968 with the following functions:

> To develop the scientific objectives; to specify the observational requirements; and to perform initial technological feasibility evaluations.
> To serve as the principal mechanism for coordination and communication between the government and the scientific community with respect to research and development relating to GARP.
> To review and advise on the detailed project design, operational logistical planning, and actual field work developed by the government.
> To serve as the U.S. link to the international scientific activities being developed for the GARP by the International Council of Scientific Unions and its member unions and scientific committees.

To study, in concert with the Committee on Atmospheric Sciences, the education and manpower needs in the atmospheric sciences nationally, and internationally in terms of potential problems that may be posed by the GARP.

Thus, the stage appears to have been set for a program which has stirred the imagination of meteorologists all over the world and one that promises to engage their attention well into the 1970's.

The incremental costs of GARP are estimated to be on the order of 100 million dollars over a five-year period. The incremental costs of the observational communications and computational aspects of the World Weather Watch are about 150 million dollars a year or, expressed in another form, an increase of about 15 percent in the cost of the present array of national systems. It is interesting to note that this figure of 15 percent would be increased several-fold were it not possible to take advantage of the advanced observational and communications capability brought within the realm of possibility by the earth-orbiting satellite. In any case, the potential savings from improved predictions over longer periods of time are quite attractive (see Part 3). It has been estimated that in the United States alone, their value is on the order of one billion dollars per year.

But more than a new technology and economic savings are at stake. The WWW and GARP place in our hands what former Assistant Secretary of State Harlan Cleveland referred to as "the technological imperatives" to create or strengthen the institutions for international cooperation that are required to serve our own national interests and that of other nations while simultaneously advancing the welfare of all mankind. Much more than the parochial promotion of a specialized area of science is at stake. Cleveland quotes President Kennedy as remarking that a manageable worthwhile system of world order will be based "not on a sudden revolution of human nature, but on a gradual evolution in human institutions—on a series of concrete actions and effective agreements which are in the interest of all concerned." If WWW and GARP add one more small brick to the edifice that restrains world conflict and supports world order, science will have served a noble purpose by enriching human life. The burden of responsibility for seeing that this happens is, I believe, on scientists. Long ago Heisenberg underscored the unique role of science in contributing to the solution of one of the great problems of our time with these words:

It is especially one feature of science which makes it more than anything else suited for establishing the first strong connection between different cultural traditions. This is the fact that the ultimate decisions about the value of a special scientific work, about what is correct or wrong in the work, do not depend on any human authority. It may sometimes take many years before one knows the solution of a problem, before one can distinguish between truth and error, but finally the questions will be decided, and the decisions are made not by any group of scientists but by nature itself.

RICHARD HALLGREN

Richard Hallgren received his B.S. degree in 1953 and his Ph.D. degree in 1960 from Pennsylvania State University. He worked for IBM as the manager of the Meteorological Systems Department until 1964, and then became Scientific adviser to the Assistant Secretary of Commerce for Science and Technology. Since November 1965 he has been the director of the Environmental Science Services Administration's Office of World Weather Systems. His research and publications have dealt with cloud physics and atmospheric electricity.

Abstract—*Whereas Dr. Malone discussed the history and future of meteorology, Dr. Hallgren focuses our attention on specific programs for international cooperation. He points out the inadequacy of present national observing networks and then proceeds to outline the development of the World Weather Watch which in its final form will be the embodiment of all the major goals of a global meteorological network.*

The World Weather Program*

MORE accurate weather forecasts for tomorrow and useful weather forecasts one week in advance are a certainty. Useful weather forcasts two weeks in advance are a definite possibility. These will be brought about by the successful execution of the World Weather Program.

It was with this intention that meteorologists from all over the world gathered at Geneva three April 1967, for the quadrennial meeting of the World Meteorological Organization. The main item of business on the agenda of this Fifth World Meteorological Congress was to chart the course of the World Weather Program. At the opening session, the leader of the U.S. Delegation, Environmental Science Services Administrator Dr. Robert M. White, received a telegram in which President Johnson instructed our delegates to "Pledge the full and continuing participation of the United States in this important endeavor."

During the past few years, a growing awareness and interest in the World Weather Program has been developing. So it should be. Its planning and execution are as much dependent upon space-age technology as they are upon meteorological science, and the results will touch, in some way, the daily pursuits of virtually everyone.

The World Weather Program has three goals: to develop a capability to make dependable long-range weather predictions (two weeks is the target), to explore theoretically the degree to which the large-scale features of weather can be modified, and to further international cooperation—without which the other goals are unattainable.

This program is vital to the United States. Each year in this country, on the average, weather disasters cost us over a thousand lives and billions

* This article first appeared in the Spring–Summer 1968 edition of the TRW *Space Log*. Permission for the use of the article was generously given by the author and by the TRW Corporation.

of dollars in loss of property. And this takes into account only those aspects of weather that make the headlines: blizzards, floods, tornadoes, hurricanes, etc. It takes little imagination to see the need for adequate advance warning of these natural calamities. But this is only a small part of the story. Many more billions of dollars are lost each year, and untold discomfort and even suffering endured, simply because of our inability to take weather factors—other than historical experience—into account far enough in the planning of agriculture, construction, transportation, merchandising, and the host of other weather-sensitive activities. To cite only one example, the consumption of natural gas is extremely variable, depending upon temperature, and it takes as long as four days for the nation's pipelines to carry the gas from producing wells in Texas to consumers in the Northern cities. The pipeline companies and gas producers use Weather Bureau forecasts and the specialized services of the best private meteorologists in a constant effort to ensure the right service to the right places at the right time. But the current state of the science of weather forecasting sometimes frustrates this intention, or at least requires the maintenance of expensive storage facilities as a hedge against unforeseen surges in demand. A miss of only a few degrees in the forecast average temperature for a large city, four or five days in advance, can cause the demand for gas to differ from the anticipated comsumption by millions of cubic feet. Similar problems of course are encountered by the producers, distributors, and consumers of other forms of energy. There will be a more complete discussion of this topic in the following article.

It would be absurd to assert that any human endeavor in the foreseeable future would totally eliminate this economic loss and human misery. But it is clear that with vastly improved weather services, a significant portion of that loss—and that suffering—could be eliminated. And the way to bring about major improvements in weather services here in the United States, as well as in all other nations of the world, is through the successful execution of the World Weather Program.

The World Weather Program recognizes the atmosphere the way it really is: a global phenomenon, not a regional one. It recognizes that no single nation can cope with the world's weather alone. And the program is timely. Since the development of the electronic computer in the late 1940's, mathematical simulation models of the atmosphere have been so successfully applied that rapid progress has been made in the understanding of the physics of the atmosphere. But scarcity of global data threatens to limit that progress. Just as the physics and mathematics of the problem grow in complexity as the scientists look further into the future, so do the data requirements. Figure 1 illustrates diagrammatically how the data required for forecasts of mid-latitude circulation patterns expand as the forecast period is lengthened. For the longer-range forcasts, data over the entire globe, from high in the atmosphere to the ocean's thermocline, are needed. Yet only 20% of the world is covered by adequate weather and marine observations. Consequently, the scientists lack observations to define, accurately, an initial state of the atmosphere and ocean for their

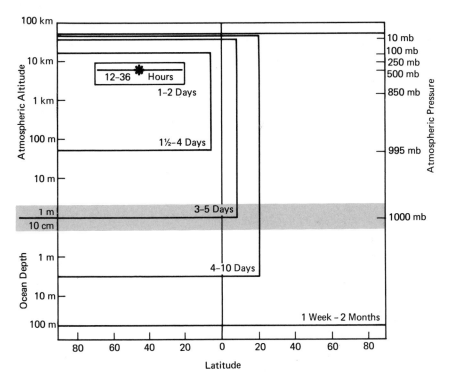

Figure 1   Schematic representation of the vertical and latitudinal data requirements for forecasts of the 500 mb (about 6 km altitude) circulation in the mid-latitudes as a function of the forecast period.

mathematical models. Equally important, observations are lacking against which to evaluate the subsequent performance of the models.

But recent progress in satellite technology promises to change that picture. With a judicious mixture of traditional observing facilities, horizontal sounding balloons, buoys, and satellite systems, it will be possible in the not-too-distant future to keep the entire atmosphere-ocean complex under regular surveillance. These developments—the computer, the satellite, and the growth in meteorological knowledge—have made atmospheric scientists confident that we can achieve a thorough understanding of large-scale atmospheric processes, create mathematical models which simulate these processes with fidelity, and hence predict their future course far beyond our current forecast range of a few days. To go one step further, this same knowledge can be applied to assessing the possibilities of large-scale weather modification. With sophisticated models, it should be possible to test theoretically in a computer whether or not weather may be controlled by artificially modifying the natural processes at those critical time and places which would trigger large-scale beneficial changes in the circulation.

In 1961 the United Nations passed two resolutions which called for the World Meteorological Organization and the International Council of Scientific Unions to develop a detailed program plan for international cooperation in meteorology. The program has since evolved into two major parts: the World Weather Watch and the Global Atmospheric Research Program, or GARP.

The World Weather Watch includes the design, development, and implementation of an international system for the regular observation of the atmosphere over the entire globe and for the rapid and efficient communication, processing, and analysis of worldwide weather data. Internationally, the planning and coordination of the Watch has been undertaken by the World Meteorological Organization—WMO.

The other portion of the World Weather Program—GARP—will be the conduct of a comprehensive program of research on the physics of the atmosphere, including major data-gathering efforts required for the research. The planning of GARP, internationally, will be done jointly by the International Council of Scientific Unions and WMO.

Three kinds of GARP data-gathering experiments are being considered. One is a series of studies of the planetary boundary—the layer immediately above the surface up to a few thousand feet. It is in this layer, through interaction with the underlying oceans, that energy is taken out or put into either system. The mechanism of this exchange must be understood so that it can be incorporated more precisely into the mathematical models which will permit long-range weather predictions.

A second group of data-gathering experiments is regional in scope. One such experiment, projected for about 1973, is in the tropics where our present knowledge is especially meager. Initial planning places this experiment—TROMEX—among the archipelagos of the Western Pacific. The third and most complex of the data-gathering experiments is a global meteorological experiment, or GLOMEX. GLOMEX is envisaged to be an intensified experiment extending over a period of up to a year to provide research scientists with a complete set of global data. Internationally, the planning has projected that experiment for the mid-1970's.

Both in the case of the Watch and the data-gathering experiments of the GARP, we need far more extensive data than are currently obtainable; in both cases we also need measurements of essentially the same parameters. At the surface, wind velocity, atmospheric temperature, humidity, pressure, amount of precipitation, and sea-surface temperature must be observed. Below the sea surface, we require the measurement of temperature and salinity down to the thermocline. Aloft, we need the distribution of clouds throughout the atmosphere, as well as the measurement of wind, humidity, temperature, and ozone at various levels. In the case of the Watch and the GLOMEX, these observations are needed on a global basis. The requirements differ only in the pecision of measurement and the spacing of observations, which are more demanding of GLOMEX, and in the lack of permanency in the case of GLOMEX facilities.

The Watch and GARP require the deployment of equipment on a global basis; therefore, both require the development and acquisition of new observational facilities. But before going into some of these new technological possibilities, let us look briefly at the present status of the two parts of the World Weather Program.

At the WMO meeting in the spring of 1967, it was agreed that it would take at least a decade to bring the full Watch into existence. The nations adopted a program for the next four years which would focus on the development of new technology for system, and on the implementation with existing technology of improvements to eliminate the more critical deficiencies in the present international weather system. In this plan, the nations of the world, including the United States, would expand the use of ships, radiosondes, and satellites for observations; communications links would be established to collect the data more rapidly, and to transmit to all nations the products of the three World Meteorological Centers which had already been established: one in Washington, one in Moscow, and one in Melbourne.

Also during 1967, the World Meteorological Organization and the International Council of Scientific Unions agreed on the composition and function of their joint group for the planning and coordination of GARP. The two organizations also sponsored a major symposium in Stockholm last summer where many of the world's leading atmospheric scientists set down the broad objectives of the three types of data-gathering experiments which were mentioned earlier. But while the international institutions are necessary, they cannot of themselves carry out the program. Much of the planning, most of the development, and all of the money for the program must come from the nations of the world.

Here in the United States, the program has received high-level attention. The President has expressed his support, and there is a joint resolution before Congress which, if enacted, will make it clear that the United States gives full support to this program. The National Academy of Sciences is deeply involved. The government has asked the academy to accept primary responsibilities for the scientific planning of the United States contribution to the international data-gathering experiments. The academy has agreed to do so, and for that purpose has organized a GARP Committee, whose membership has been selected from among the nation's leading atmospheric scientists.

Within the government, the broad plan for United States participation in the World Weather program was established by a group appointed in 1964 by the Secretary of commerce at the request of the President. The President has since named the Environmental Science Services Administration—ESSA—in the Department of Commerce to be the lead agency in the government for the overall coordination and planning of the U.S. contribution to the program. ESSA discharges this role through its Office of World Weather Systems. The other agencies with primary roles in the World Weather Program are the National Science Foundation (support of GARP research and field experiments and the training of scientists), National Aeronautics and Space Administration—NASA— (development of space technology), Department of State (foreign policies, guidance and support of assistance programs for the developing nations), and the Department of Transportation (support of operations in data-gathering experiments and development of buoys).

Concerning the technological developments critical to the World Weather Program, it should be made clear that immediate improvements in the world's observing network can be made—and are being made—by expanding conventional observing facilities. However, to create a complete global system purely on this basis would be very costly. The new technology offers promise of reducing the cost substantially.

Much of the new technology is based on exploiting more fully the potential of satellites. Today, ESSA's satellites—the world's first operational satellite system—are providing daily pictures of global cloud cover. Two types of television camera systems on ESSA satellites in polar orbits are currently in operation. One of these systems continuously transmits its field of view during daylight hours to any receiving station within line-of-sight range. Over 300 of these Automatic Picture Transmission

(APT) receiving sets are in use today in some 50 countries. Figure 2 shows a typical ATP receiving facility. The other operational ESSA optical system stores the television picture data for each orbit on magnetic tape for bulk transmission to appropriately located ground command-control facilities. From these stations the data are relayed to ESSA's National Environmental Satellite Center. There, with the aid of a computer, cloud mosaics are constructed to show the cloud distribution over major parts

Figure 2   A typical APT facility for local reception of cloud pictures from the ESSA satellite.

of the globe. Figure 3 depicts ESSA 3, which provides the type of cloud data just described. Soon, ESSA satellites will mount infrared radiometers to reveal cloud patterns by night.

Cloud photographs are also being taken by cameras on NASA's synchronous satellites ATS 1 and 2. They have proved to be valuable for providing information concerning the complex features of the tropics, and in detecting cloud motions which provide information on global circulation patterns and storm movements. Figure 4 is an example of the kind of pictures obtained from ATS 1.

However, cloud pictures, though very useful, do not meet our need for global data. Before long-range forecasts are possible, we must make quantitative measurements of physical parameters. The satellite, with its unique capability for global surveillance, provides a vehicle for acquiring much of the data required.

The function of the satellite in meteorological observations may be divided into two broad areas. In one area, the satellite is the medium for

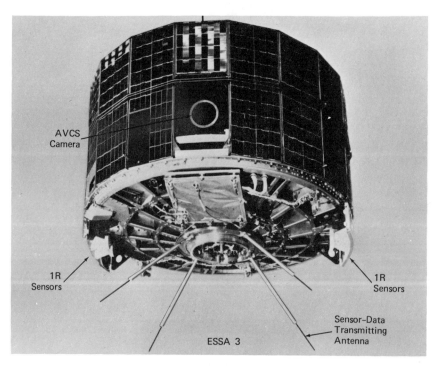

Figure 3   ESSA satellite equipped with IR radiometer and advanced vidicon camera system (AVCS) which stores global cloud pictures for readout once each orbit.

data location and collection from a variety of platforms in the atmosphere or on the earth's surface. In these systems, direct sensors are attached to the terrestrial platforms and their data are telemetered to a satellite upon interrogation. Figure 5 illustrates this concept. In the second area, the satellite itself serves as the platform for instruments which take remote measurements from which atmospheric parameters may be derived.

Some of the more interesting developments in current meteorological technology are to be found in the work being done with sensor platforms to be located and/or interrogated by a suitably equipped satellite. The National Center for Atmospheric Research in Boulder, Colorado, is engaged in the development of "superpressure" or "horizontal sounding" balloons which will float with the wind currents at constant density levels in the atmosphere and serve as upper-air sensor platforms. These balloons are made of a nonextensible material, such as some of the polyester films,

Figure 4 Hemispheric cloud patterns as seen from NASA's ATS 1.

and are filled with enough lifting gas to create a small positive inside pressure ("superpressure") at design altitude. The superpressure keeps the balloon size and shape relatively constant so that it will rise to a predetermined density level and float there. Some of these balloons have floated round and round the world for close to a year. Figure 6 illustrates the three months' track of one such balloon in the Southern Hemisphere. The balloons will carry sensors for pressure, temperature, and water vapor,

Figure 5   Artist's concept of the weather observation system of the future. Satellites aloft are relaying data gathered by the buoy, balloons, land stations, and ships shown below. Other satellites are taking cloud photographs and probing the atmosphere with spectrometers.

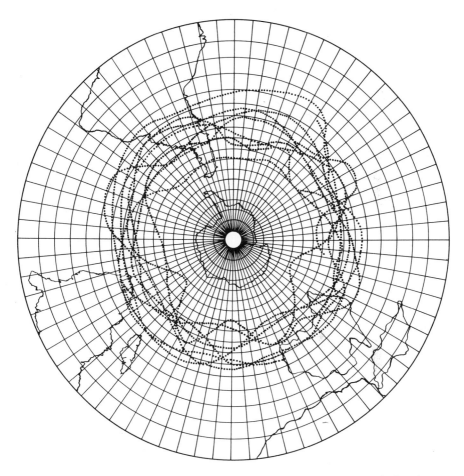

Figure 6   The path followed by a superpressure balloon at 200 mb for the first three months of its life. The balloon was launched and tracked by a team from the National Center for Atmospheric Research from Christchurch, New Zealand. This balloon stayed aloft and its electronics package functioned for nearly a year.

and a transponder to answer interrogation signals (see Dr. Ruttenberg's discussion of the GHOST System in Part 7).

Another sensor platform, to which a great deal of attention has been directed recently, is the ocean buoy for the measurement and telemetering of oceanographic and surface meterological data. Despite the long history of buoy design, test, and operation for a great many purposes, we have yet to achieve a buoy system—hull, power supply, sensors, and telemetering equipment—which meets the performance requirements of both the oceanographer and meterologist while yet being economical to deploy, service, and operate on a global basis. Within the government, the Coast Guard now has the responsibility for the design and development of such buoys. Figure 7 is an illustration of an early specialized meteorological

Figure 7   U.S. Navy's NOMAD buoy, anchored in the Gulf of Mexico. Regular weather observations are telemetered to land station over 300 mi away.

buoy design. While designed and built many years ago, the pictured Navy NOMAD is still used for telemetering limited meteorological and oceanographic data from an anchorage in the Gulf of Mexico.

Automatic sensing and telemetering equipment suitable for placing on merchant ships and similar land-situation installations are the other platform being considered for use in satellite data location and collection systems for the World Weather Program. To operate with these various direct-sensing platforms, NASA has under development two interrogation and locating systems, and the French Centre Nationale d'Études Spatiales is actively engaged in developing a third, the EOLE experiment, as described in Part 7 by Dr. Morel.

The three systems are all designed to interrogate the platforms, locate their positions, and receive the sensor data for subsequent relay to ground command-control stations. The three systems differ in their techniques for locating the platform's position. One NASA system uses the Navy's OMEGA navigational system to locate the platforms by including OMEGA receivers in the platform telemetering devices. The other NASA system and EOLE use different ranging techniques for establishing the platform's position relative to known positions of the satellite. When used with super-pressure horizontal sounding balloons, these systems not only can locate the position of a given observation, but can also observe the winds at the balloon flight level by determining the change in location of a given balloon between interrogations. NASA scientists carried out the initial tests of these systems for ATS C and D spacecraft during 1968 and Nimbus B-2 in 1969. (Nimbus B was destroyed during an unsuccessful launch attempt in May of 1968.) More complete tests will be conducted in 1970 on Nimbus D, and the French expect to conduct the EOLE experiment during the same year. By the end of 1970, it should be possible to ascertain which of the three systems is best suited for the operational World Weather Watch and for GARP experiments. Technologically, there is little doubt that at least one of the direct-sensor satellite systems just discussed can be perfected.

In the satellite systems of the second major category, those involving indirect sensors mounted on the satellite itself, the satellite-borne meteorological instruments are all intended to sense atmospheric behavior with respect to radiation in some portion of the electromagnetic spectrum. The cloud pictures obtained by two such remote sensors—the TV camera in the visible wavelengths and the infrared radiometer—have already been mentioned.

In the infrared, satellite spectrometers and interferometers for measuring the vertical distribution of temperature and water vapor in the atmosphere are well along in their development by ESSA and NASA. The forthcoming Nimbus B-2 and D programs (see Figure 8) include the first satellite flight tests of these very important sensors. The feasibility of the technique has already been proved by comparing data derived from a balloon-borne IR spectrometer, SIRS (Satellite Infra-Red Spectrometer), with conventional radiosonde observations. Figure 9 is an example of one of these comparisons. SIRS senses radiation in seven very narrow spectral bands—six in the 15μ CO_2 absorption band and the seventh in the 11.1-μ window. These observations permit the derivation of temperature as a function of pressure at six levels in the atmosphere. It is estimated that the derived sounding will approximate that obtained by a radiosonde ascent within 1 or 2°C. The mathematics of interpreting the radiation measurements in terms of vertical temperature distribution are exceedingly complex, but timely solutions are possible with the modern computer. When used in conjunction with the temperature data derived from SIRS, the distribution of water vapor may be inferred from IR radiation measurements taken by other satellite-borne sensors.

Figure 8   Representation of a NIMBUS satellite in orbit. NASA uses NIMBUS spacecraft as test beds for new meteorological instruments and satellite systems in the development stages.

Ozone profiles may be similarly deduced by ozone's ability to absorb certain ultra-violet wavelengths. The backscatter ultra-violet spectrometer (or BUV) is being developed for that application on satellites. The BUV will also be flight tested in the NIMBUS program.

Another application in the visible spectrum is the star-occultation technique, which is in the early stages of development by NASA. In that system the index of refraction, through portions of the atmosphere, will be measured by noting the exact position of an orbiting satellite at the time of disappearance of selected stars over the satellite's horizon. Since the index of refraction is related to air density, this technique offers the possibility of deducing the approximate distribution of temperature and pressure through slices of the atmosphere (see Mr. O'Brien's article in Part 7).

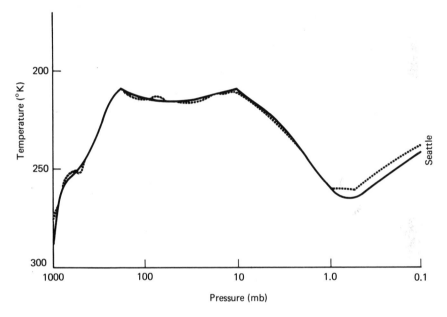

Figure 9   Comparison between atmospheric temperatures sounding by remote IR spectrometer (solid line) and by a standard Weather Bureau radiosonde (dotted line).

The vertical profile of temperature and moisture can also theoretically be derived from passive spectrometric measurements at microwave frequencies. Such observations from a satellite offer the possibility of going all the way to the surface even in cloudy weather, since selected microwave missions can penetrate through clouds. Another microwave technique being considered by NASA advanced planners involves the measurement of the refraction of microwaves between widely spaced satellites in the same orbit. As with the star occultation already mentioned, these changes in refraction can be related to air densities. But at present, progress in the microwave area is limited due to the lack of small, lightweight components that consume little power. Hopefully, it is only a matter of time until physicists and electrical engineers change that situation.

Very recently, the possibility that the backscatter from laser beams could be used to measure meteorological parameters has received some attention, but there again the technology of lasers must be further developed before a meteorological laser system can be applied to the satellite (see Dr. Ligda's article in Part 7).

These are only some of the many possibilities that exist for remote sensing of the atmosphere from satellite-borne instruments. But it is enough to illustrate the point that the World Weather Program offers almost unlimited possibilities for applying new technological advances. We are still depending upon the ingenuity and insight of the nation's technologists to show us how we can obtain global weather observations with the accuracies and resolutions we require, at costs the world's weather services can afford. Much has been done in that respect already, but there is still plenty of room for new ideas.

ROBERT ELLIOTT

Robert Elliott studied physics as an undergraduate and meteorology for a M.S. degree at the California Institute of Technology. After graduating in 1937, he worked for two years as a meteorologist for Eastern Airlines and then returned to the California Institute of Technology to teach and do research in extended-range weather forecasting. After World War II, when he was an officer in the Navy's Long Range Forecasting Unit, he returned to teaching. In 1947, he became Director of Research for the American Institute of Aerological Research in Pasadena, California and in 1955, the President of North American Weather Consultants in Santa Barbara, California. Mr. Elliott has written numerous articles on cloud seeding and weather modifications as well as on aspects of long-range forecasting.

Abstract—With his exerience as a weather consultant and forecaster, Mr. Elliott concludes this chapter on goals of a global weather system with a look at the present as well as the future of weather prediction. He describes one of the current ways of predicting future states of the weather, the analog method, and goes on to discuss the stage beyond prediction, which is weather modification.

The Analog
and Weather-Type Methods
of Weather Forecasting

THE topic of this article is the analog and weather-type methods of weather forecasting as related to current developments in meteorology. In analog forecasting, past weather maps which closely resemble the current weather map are selected from the historical map file, and the weather developments from these cases are used as a basis for making weather predictions. The basic elements are the flow patterns as represented by isobars, pressure contours, and fronts, and weather elements such as cloud cover, precipitation, and temperature. For example, a forecaster for the New York area might look back in his file of weather maps to find several with pressure patterns similar to those currently observed. Then by noting where in each case an approaching front moved and what rainfall occurred, he could predict with a certain probability that rain would fall in the New York area during the ensuing forecast period.

In matching the current map to past maps the first question to be answered is: Over how large an area should the comparison be made? It turns out that the optimal area increases with the length of the forecast For a two- to three-day prediction on area of about 45° longitude by 50° latitude is satisfactory. The region from 20° to 70° north embraces the principle storm zone in the Northern Hemisphere so that this is the preferred latitude range. The longitudinal sector should lie mostly upwind of the point for which the prediction is to be made.

Of course, the "eyeball" comparison and selection of analogous weather maps is crude, and one naturally turns to more objective methods. Thus, weather-map pressure patterns may be correlated by statistical methods, those with the highest correlation being selected as analogs. A more advanced procedure is to express the pressure pattern of each map in terms of the amplitudes of orthogonal polynomials. The best matches between the principle terms of historical maps and the current map are then the analogs.

What has been described thus far can be termed a "thumbprint" analog. Suppose that instead of matching just single historical days with the current day, one matches blocks of, say, three sequential days from the historical file, with the block of three days terminating with the current day? We should then be selecting analogs based upon matches over a period of time and over a geographical area. It was found years ago that this approach led to weather predictions that were more accurate than those provided by thumbprint analogs if "extended"-period predictions were the desired end product.

As might be expected, the matching of patterns through sequences of days on an objective numerical basis is quite difficult. During World War II, before present computing facilities were available, it was useful to compile catalogues of historical "weather types" covering blocks of three days, in one variant, and of six days in another. The number of weather types was ten, but there were numerous subtypes in the different seasons. The members of one type could be considered analogs. It was possible to look into the catalogue of weather types and make up what is called a synoptic climatology, that is, a set of statistics on the probability of rain, warm or cold temperatures, or other weather events at specific localities such as San Francisco or New York for each day of each weather-type sequence.

A sequence of hypothetical or "ideal" weather maps was often employed to represent the weather type for matching purposes. Figure 1 shows such a sequence for type A (six-day type): winter. A southeastward movement of storm fronts into the Great Basin area followed by an east-northeastward motion of lows across the Mississippi and Ohio valleys is characteristic of type A. Two main systems move across the country during this type of weather, the second of which develops into a major storm in the East. During the six-day period precipitation averages well above normal in the Mississippi and Ohio valleys. Temperatures are below normal in the West and above normal in the East. The surface isobars and surface frontal positions are shown in Figure 1. A characteristic flow pattern aloft is associated with each weather type (not shown in the figure). This upper pattern tends to be stable during the course of one type of weather, but may change abruptly as a new type of weather moves into the sector. Storm centers are "steered" along the upper-level flow pattern. It is possible to represent in a meaningful way the essential features of a given weather type on a single chart by focusing attention on this stable upper-level pattern.

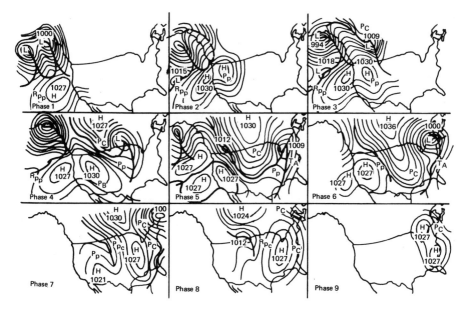

Figure 1 Ideal weather maps over the North American continent for weather type A.

The ten charts shown in Figure 2 represent in schematic form the essential features of the ten designated weather types. The heavy solid lines are streamlines of the flow pattern at the 500-mb level (18,000 to 20,000 ft) and indicate approximately the track of storm centers. The cross-hatched and spotted areas represent regions of persisten⊾ ㅤsurface high or low pressure, respectively. The double arrows lie in regions where cold polar outbreaks follow the eastward passage of cyclonic storm systems. For example, the sketch for type A shows the cold outbreak coming southward over the far western states, thus accounting for below-normal temperatures there. The upper flow is more southerly to the east of the Rockies, accounting for the warmth there.

A very different type is *Bn-b,* in which the cold air moves down just east of the Rockies producing a "cold wave" in the Great Plains states. The persistent high in the West shelters the southwest coast from Pacific storms, which move into Washington and British Columbia. A persistence of *Bn-b,* and of *Bn-a,* in the winter season leads to drought conditions in the Southwest.

The weather types appearing in the left-hand column of Figure 2 have well-developed "waves" in the upper-level flow pattern. This results in

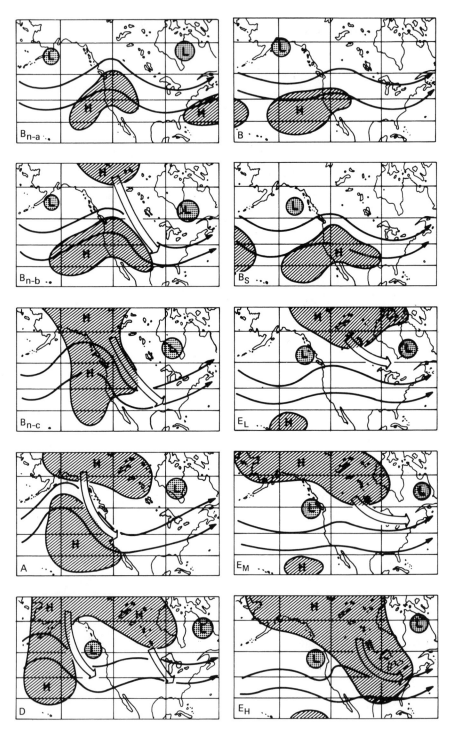

Figure 2   Schematic diagram of the ten basic weather types for North America showing flow patterns at the 500-mb level.

a great deal of cross-latitude exchange of cold- and warm-air masses. Hence such weather types are called meridional flow types. The types appearing in the right-hand column have a much flatter wave form and are called zonal flow types. Jet streams are most fully developed in zonal flow types. Note that each successive type in the left-hand column shows a more westward position of the dominant ridge line. Each successive type in the right-hand column shows a more southerly mean position of the upper flow pattern.

The archive of some 65 years of historical weather maps has been catalogued in terms of weather types. In making use of these in the preparation of extended-range weather forecasts, the statistical probabilities, computed from historical weather data, can be used as an aid in predicting the weather, once the current weather type is established. This approach has already been indicated above. In addition, the catalogue can be employed as a means for sorting out specific cases bearing an unusual resemblance to the current sequence; that is, particular analogs can be selected from the set of corresponding weather-type cases. This approach was used most extensively during World War II.

After the war there was a lapse in interest in the analog approach to weather forecasting as the concept of numerical prediction came into view. However, the atmosphere is a very complex body of energy sources and sinks, and it is difficult to introduce enough parameters into a model to predict its behavior over any appreciable length of time. The advantage of the analog method is that it uses the past history of the atmosphere as a model and so can incorporate all its complexities into weather predictions.

The advantages of the purely numerical model of the atmosphere as handled by some very large, fast computers have been considerable, and a large amount of progress has been made in this direction (as discussed by Dr. Leith and Dr. Mintz in a later article). However, there are signs of renewed interest in the use of the analog and weather-type methods to handle the problem of longer-range forecasts. Now that the numerical tools and statistical procedures are available it should be possible to make rapid progress in improving the objectivity and convenience of such methods.

A step beyond using weather types for analog selection is the investigation of certain sequences of types which appear to be associated with what is known as the "index cycle." There is a tendency for zonal weather types to persist for a period of about a week followed by a breakdown into meridional types which become more extreme as large stagnant "blocking" high-pressure areas develop. Over a period of two weeks everything swings back to the zonal type. Although these are not precise cycles, it is conceivable that we could assign numbers to represent the degree of zonal flow, the degree of meridional flow, and the degree of blocking appearing within a weather type, and then work out a linear prediction system that will forecast the degree of zonal flow, meridional flow, and blocking flow many days in advance. If we can predict these three numbers, then we can predict, on a probability basis at least, future weather types, and from them the weather. Statistical prediction methods of this sort could be developed from the historical weather-type catalogues themselves, but it would be better first to redefine the weather types in objective numerical terms.

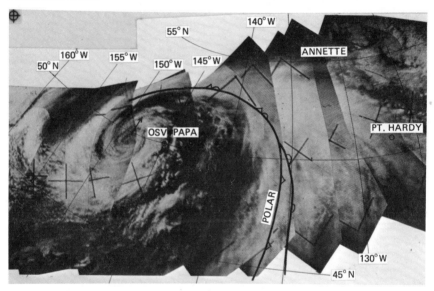

Figure 3   Satellite photograph of cyclone and weather front.

In the last five years, since TIROS and Nimbus have come of age, an increasing number of high-quality satellite weather pictures have become available to meteorologists. The question of how to make best use of these pictures has not been fully answered. Figure 3 is an example of such a picture. The spiral cloud system is a fully occluded cyclone with a deep low-pressure center located at the center of the spiral and a frontal system extending out of the center toward the southwest, then the west. Most of the storm activity is associated with the several bright areas indicating deep clouds in which there is a high likelihood of precipitation. Although this picture contains one complete cyclone and associated front, it is a composite of several separate satellite photos cut and glued together. Such mosaics can be made to cover areas as large as weather maps by putting photos together from successive passes. Now it should be possible to collect a historical file of these composite pictures and to compare these with current satellite pictures in order to identify analogs. This would probably give results as significant as those obtained by matching pressure patterns. Over ocean areas, where synoptic reports are scarce, these analogs might be more significant.

One of the major problems in making use of the data from satellite cloud pictures is that it is difficult to correlate them directly with weather maps based on pressure patterns. The satellites take direct pictures of the cloud conditions, but since it has not been possible to obtain any information about pressures or densities, the pictures contain very little numerical information. Accordingly, it would appear that their direct use in analog and weather-type methods would at present be the most feasible way to make use of the wealth of weather information available in them.

Eventually, of course, there will have to be ways of obtaining pressure-versus-height data directly from remote sensing devices on board satellites. When this information is available, the satellite pictures will be capable of supplying data directly to the numerical as well as to the analog weather systems.

Accurate long-range weather prediction is clearly a major goal of any global weather forecasting system. However, as the field of weather modification grows, there is an increasing interest in the possibilities of influencing weather on a sizable scale. Once we understand in detail how weather systems are born and move it should be possible, not only to predict with accuracy their behavior, but to speed up or slow down the reactions which supply the enormous quantities of energy required for a major storm system. It is interesting to contemplate the scale of the energies involved. The power required to drive a full-grown cyclone is about 3×10^{10} kW acting over a three- or four-day period. Compare this power of a single storm to the total U.S. power production, which is currently about 3×10^{8} kW. A thunderstorm and a tornado, on the other hand, expend energy at a rate of about 10^{8} kW for a period of several hours and 15 minutes, respectively. They are of an equal intensity, but the tornado has a much shorter lifetime and is concentrated in a much smaller volume.

It would at first sight seem hopeless for man even to think of producing significant changes in the intensity or motions of such systems in view of their enormous energies. But there are ways in which small inputs of man-made energy can trigger the release of larger natural energies available in latent form. Thus, man can to some extent select the time and places for triggering large energy releases. Consider, for example, the silver iodide smoke generator employed for seeding clouds to induce, or increase precipitation. Its principle is to overcome the natural dearth of ice-forming nuclei artificially by supplying large numbers of submicron-size silver iodide smoke crystals. Subcooled cloud droplets, too small to have an appreciable fall velocity, are then converted to ice crystals and grow to snowflake size and descend. The point is that a small smoke generator can produce enough crystals to seed a cubic mile of cloud in less than a minute. The heat energy released in connection with the change in phase is huge, and even though only a few percent of this is converted into the kinetic energy of larger-scale motions, nevertheless the power production resulting from seeding an area of several hundred square miles can be 10^7 to 10^8 kW. Accordingly, we should not be discouraged about the possibility of manipulating weather on a grand scale.

It is interesting to note that some of the methods employed for evaluating the effects of seeding to increase precipitation in watershed areas are similar to analog methods in that seeding-produced alterations in historical precipitation pattern relationships are looked for. The search, however, is for dissimilarities rather than for similarities.

PART 3

benefits

JACK C. THOMPSON

Jack C. Thompson was with the U.S. Weather Bureau for more than 35 years. He has engaged in considerable research on the economic utility of weather predictions, and is the author of a number of technical papers on the subject. In 1963, he served as a consultant on meteorological satellite problems to the World Meteorological Organization, and recently was appointed by that agency as chairman of a special working group concerned with the economics of global meteorological services. Since the autumn of 1965, Mr. Thompson has been on the faculty of San Jose State College in the Department of Meteorology.

Abstract—*As in any other major public work, the advocates of a global weather forecasting network must present cogent evidence to government officials and ultimately to the taxpayers that a sizeable investment in such a program is worthwhile. Is it possible, however, to attach a monetary value to a better weather forecast? How does one evaluate the savings involved in long-range hurricane warnings? In attempting to answer questions such as these, Mr. Thompson proposes a language and a set of mathematical criteria with which to estimate the economic value of increased accuracy in weather predictions. As an example, he calculates values associated with certain potential improvements in three sample cities: Washington, D.C., Salt Lake City, and San Francisco.*

Costs and Benefits of Weather Prediction*

DURING recent years, advancement in meteorological technology have greatly increased our ability to observe, analyze, and understand our atmospheric environment. Eventually, such developments will make it possible for the meteorologist to provide accurate long-range weather forecasts and other improved weather services to the public. However, it seems evident that such improvements will require the expenditure of a great deal of time, effort, and material resources in order to attain these goals. The necessary hardware alone, meteorological satellites and high-speed computers, for example, are very expensive.

To some extent, such large costs may be justified by the increased scientific knowledge obtained. However, in this practical world, those who are asked to pay the cost of large meteorological projects may also, quite reasonably, be expected to ask questions about the potential economic, social, or other benefits which may be anticipated. It is the purpose of this discussion to outline a general, if preliminary, approach to this problem.

To begin with, it is necessary to examine briefly the general nature of weather information. To anyone who has planned a picnic on the basis of a fair-weather forecast only to find the outing marred by a day of rain, it hardly seems necessary to point out that a certain amount of uncertainty is associated with weather predictions. However, the results of this uncertainty are basic to an evaluation of weather data. The uncertainty arises partly because the techniques used to prove the atmosphere are handicapped by a technological inability to obtain an exact formulation or solution of the prediction problem. Although some improvement in this situation may be expected as a result of better observing and forecast-

* Based on J. C. Thompson, "Economic Gains for Scientific Advances and Operational Improvements in Meteorological Prediction," *Journal of Applied Meteorology,* Vol. 1, No. 1, March 1962, pp. 13–17. (Reprinted here with permission.)

ing methods, the basic characteristics of uncertainty will almost surely continue to be operationally significant for the foreseeable future. It is therefore necessary to any objective assessment of economic or other gains in meteorology to account, in a quantitative fashion, for this inherent uncertainty in weather information.

To accomplish this objective, the following terms are defined:

G_m = total expected gain for $N = (f_w + f_{nw})$ occasions if protective measures are taken on each occasion.

G_{nm} = total expected gain for N occasions if no protective measures are taken.

C = cost of protection on each occasion that protective measures are taken.

L = loss suffered on each occasion that adverse weather occurs and no protective measures have been taken.

X = average gross profit for each occasion, exclusive of the cost of protection C or the loss L that may have resulted.

f_w = frequency of adverse weather.

f_{nw} = frequency of favorable (nonadverse) weather.

Now, if protective measures are taken on each occasion, the total gain will be the average profit minus the cost of protection, both multiplied by N, the number of occasions. Thus,

$$G_m = (X - C)N \qquad (1)$$

Similarly, if no protective measures are taken, it will be seen that

$$G_{nm} = (X - L)f_w + Xf_{nw} \qquad (2)$$

If one desires to maximize the total gain over the entire period of a repetitive operation, protective measures should be taken whenever $G_m > G_{nm}$, or from Equations (1) and (2),

$$(S - C)N > (X - L)f_w + Xf_{nw} \qquad (3)$$

With a little algebra, this reduces to

$$f_{w/N} > \frac{C}{L} \qquad (4)$$

The left side of the expression defines P, the "probability" of adverse weather. In a similar manner it may be shown that protective measures should not be taken whenever $P < C/L$, and either course may be followed if $P = C/L$. Thus, a criterion for making the decision to protect or not protect may be written

$$P \begin{array}{c} > \\ = \\ < \end{array} \frac{C}{L} \begin{cases} \text{protect} \\ \text{either course} \\ \text{do not protect} \end{cases} \quad (5)$$

The value $P = C/L$ therefore is a critical ratio, above which protection should be provided, and below which it should not. It is interesting to note that for C, L, and X as defined here, X drops out and need not be considered in making the decision. Alternative, but generally more complex, expressions may be derived by defining these terms in a different manner.*

If, now, a series of N probability weather predictions are made, the results may be presented as shown in Table 1. Here W and No W are the occurrence and nonoccurrence, respectively, of an operationally adverse weather event, and a, b, c, and d represent the frequencies in the indicated boxes in the table.

TABLE 1

*GENERALIZED CONTINGENCY TABLE SHOWING RESULTS
OF PROBABILITY PREDICTIONS*

Observed Weather	Forecast Probability $P \leq C/L$	$P > C/L$	Totals
No W	a	b	$a + b$
W	c	d	$c + d$
Totals	$a + c$	$b + d$	$N = a + b + c + d$

Assuming the criterion of Equation (5) has been used, a series of optimum decisions will have been made. From the table, then, the total weather protection expense for the operation E_f will be due to the cost of protection whenever protective measures have been taken ($P \geq C/L$), plus the loss suffered whenever no protective measures have been taken ($P \leq C/L$) and adverse weather W occurs. Thus,

$$E_f = C(b + d) + Lc \quad (6)$$

On the other hand, if it were scientifically possible to make "perfect" forecasts, the total expense for the operation E_p would arise only from

* See, for example, I. I. Gringorton, "Forecasting by Statistical Inferences," *Journal of Meteorology,* Vol. 7, No. 6, December 1950, pp. 388–394.

the necessity for protecting against adverse weather. Thus,

$$E_p = C(c + d) \tag{7}$$

Now to obtain the economic gain which might be achieved if perfect forecasts were attainable, exceeding the value currently obtainable within the prevailing state of the science, take $E_f - E_p$. However, it is convenient to present this information in "nondimensional" form, that is, as the gain per unit forecast, per unit of loss. Thus, the potential gain for a millenial improvement in scientific knowledge G_s would be

$$G_s = \frac{E_f - E_p}{NL} = \frac{1}{N}\left[\frac{(b - c)C}{L} + c\right] \tag{8}$$

Now of course current weather forecasts do not usually contain quantitative information about their uncertainty, so that the value of P is not normally available. Instead, a working assumption, more or less equivalent to an "average" value of the assumed economic risks is used to produce a categorical prediction of the future weather. The results of a series of such predictions may again be presented as in the contingency table, but with the ratio C/L assumed constant. The expense for these operation decisions E_a will be given by

$$E_a = C(b_a + d_a) + Lc_a \tag{9}$$

where the subscript a denotes a categorical prediction made at a fixed decision level.

The economic gain which could be realized from an optimum use of uncertainty information, exceeding the value of these categorical "average" predictions, is then given by the difference between Equations (6) and (9). Denoting this improvement as G_o, and presenting the information as before, gives

$$G_o = \frac{E_a - E_f}{NL} = \frac{1}{N}\left[\frac{(b_a + d_a - b - d)C}{L} + c_a - c\right] \tag{10}$$

Finally, it is of interest to obtain the total economic gain which would be realized if perfect forecasts were attainable, exceeding the value of currently issued average predictions. This is given by the difference between Equations (9) and (7), or alternatively, by the sum of Equations (8) and (10). Denoting this total economic gain as G_t gives

$$G_t = \frac{E_a - E_p}{NL} = \frac{1}{N}\left[\frac{(b_a + d_a - c - c)C}{L} + c_a\right] \tag{11}$$

It is now of interest to apply Equations (8), (10), and (11) to weather predictions. For this purpose, three series of probability forecasts have been analyzed and the results are presented in Figures 1 to 3. In each case, the economic gain G is the vertical coordinate, and the operational risk ratio C/L is the horizontal coordinate. Figure 1 shows an analysis of 24-hour temperature forecasts which were made in Washington, D.C. Figure 2 contains a similar set of predictions for Salt Lake City, Utah;

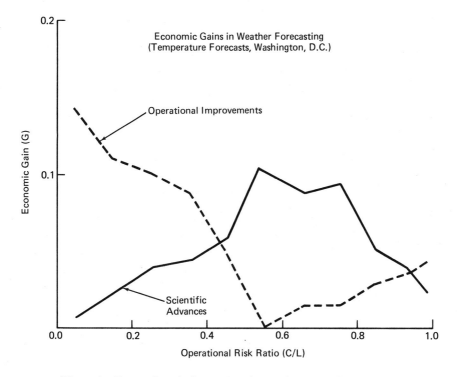

Figure 1   Economic gain in weather forecasting—Washington D.C.

and Figure 3 is a third sample of 24-hour rainfall forecasts for San Francisco, California. The similarity of the three sets of data, both in general appearance and in magnitude, is obvious. In general, the maximum gain for "perfect" forecasting is of the order of 10 percent, and, although the greatest value of scientific advances would be achieved for operations in which the risk ratio C/L is near the middle of the range, the operational improvements would be greatest for these activities with a risk ratio near the extremes of the range.

It is also of interest to determine the average economic gain from these analyses. If it is assumed that the operations represented by the horizontal scale are equally likely and equally important, a simple arithmetic mean of the economic gain for each curve can be computed. Such mean values are shown in Table 2.

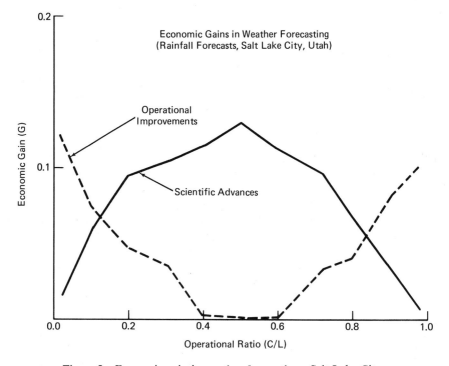

Figure 2   Economic gain in weather forecasting—Salt Lake City.

Again, for all three cases the values are strikingly similar. Now, if one were to assume that these percentages were reasonably general, an estimate of the potential benefits to be gained from operational and scientific improvements in weather prediction could be obtained by applying them to values of the protectable portion of current losses due to adverse weather. An initial attempt to obtain an "order-of-magnitude" estimate of such losses is available in an unpublished report to the Department of Commerce.*

According to this report, in the United States, the total annual losses

* M. E. Senko, "Weather Satellite Study—A Special Report," U.S. Department of Commerce, Weather Bureau, Washington, D.C. (unpublished).

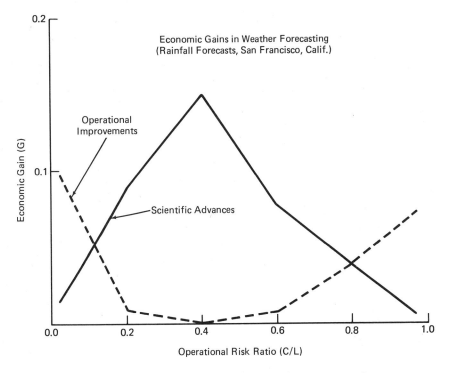

Figure 3   Economic gain in weather forecasting—San Francisco.

due to weather are on the order of $10 billion. Of these losses, about four-fifths are due to catastrophic weather—tornadoes, hurricanes, and the like—or are the result of other phenomena against which no protective measures are practicable. The other one-fifth, or about $2 billion, represents losses which improved weather information could alleviate. Applying the average percentages obtained from the previous discussion (Table 2)

TABLE 2
MEAN VALUES OF ECONOMIC GAINS FROM SCIENTIFIC
ADVANCES G_s, OPERATIONAL IMPROVEMENTS G_o, AND
TOTAL GAINS G_t

	G_s	G_o	G_s
Temperature forecasts (Washington)	0.05	0.06	0.11
Precipitation forecasts (Salt Lake City)	0.05	0.04	0.09
Precipitation forecasts (San Francisco)	0.06	0.04	0.10

Note: Figures are in units of potential loss L per forecast.

to this figure, one obtains for potential scientific improvements 0.05 × $2 billion, or $100 million; for potential operational improvements 0.05 × $2 billion, or another $100 million. Adding these two values gives an order-of-magnitude sum of $200 million for the total potential economic benefits which might accrue to the economy of the United States as a result of improved weather forecasts for periods of approximately 24 hours.

At this point a *caveat* is clearly in order. These values have been obtained through the use of (1) rather simple, albeit quite general, decision theory, (2) applied to a limited sample of meteorological data, and (3) using order-of-magnitude and somewhat incomplete estimates of weather losses in the United States. Quite evidently, the resulting values can be no more accurate than the assumptions and the data on which they are based.

As a matter of fact, the dollar estimates are presented here primarily as a means of illustrating a possible method for attacking, in a preliminary fashion, a difficult and complex problem. Further work in this area is clearly required.

Some idea of the nature of the complexities involved may be obtained from studies which have to be conducted for the raisin industry in California.* Improved methods of weather decision making in the San Joaquin Valley would, from initial considerations, increase the annual value of the crop by nearly $10 million. However, further examination of the problem reveals a lack of demand for raisins significantly beyond present production. Accordingly, if the supply were increased, the price might well decrease and some acreage would be forced out of production. Assuming further, however, that the land thus released would be used for production of other fruits, for example, almonds, cherries, and peaches, for which an increasing demand exists, it is concluded that an additional $22 to $45 million annually might be realized.

It should be recognized, therefore, that the problem of evaluating improved weather services is beset by many interacting complexities and that some of these may produce variations which could exceed the initial estimate obtained by ignoring their effects. It seems clear that this problem offers a challenging and useful field for interdisciplinary study involving the economist, the meteorologist, and weather-service users in many activities.

---

* See L. L. Kolb and R. R. Rapp, "The Utility of Weather Forecasts to the Raisin Industry," *Journal of Applied Meteorology,* Vol. 1, No. 1, March 1962, pp. 8–12; also L. B. Lave, "The Value of Better Weather Information to the Raisin Industry," *Econometrica,* Vol. 31, No. 1–2, January–April 1963, pp. 151–164.

Abstract—*This article* contains the results of studies made by the Stanford Space Systems Engineering course on the economic benefits which may result from a global weather prediction system such as that proposed in the SPINMAP report. The object of course is to decide whether the added costs of the global system will be matched or exceeded by observable benefits to various segments of the world population.*

This article complements Mr. Thompson's discussion of a theoretical decision-making criterion by presenting the results of a study of several key industries. This was an attempt to understand how and when unexpected adverse weather might affect production schedules, fuel supplies, construction operations, times of harvesting crops, and ocean routes; and finally to determine what part of the losses involved might be avoided by better and longer-range weather predictions.

* This article was written by a group of people.

Benefit Analysis
of a Global Weather
Prediction Network*

A S Mr. Thompson pointed out in the previous article, it is very difficult to arrive at an accurate estimate of the incremental benefits in a certain industry due to better weather predictions. The decision criteria are still largely subjective. Thus in this article, only the general economic and social factors which must be taken into account in making the best possible estimate are presented. Those interested in specific figures are referred to the SPINMAP report and to the references contained therein.

No comprehensive study of all the benefits of improved weather forecasts seems to have been attempted. Faced with the problem of assessing who might benefit, and to what extent, we decided to assess those industries which might be the principal beneficiaries. This article contains discussions of agriculture and air and marine transportation. The SPINMAP report contains analyses of construction, fuel, insurance, marketing, and benefits to government as well. The benefits to government were considered since most governments provide or subsidize a considerable amount of replacement cost for destruction caused by weather.

Three techniques were used to estimate the benefits to be expected from implementing the system proposed in the SPINMAP report. A survey was made of the ever-increasing literature on economic benefits of improved forecasts. Although this proved to be valuable background information, the majority of the studies to date are only qualitative and thus are unable to help assess specific benefits. Secondly, both objective and subjective assessments of the benefits in a few select sectors of the economy were made. In this, previous studies of the raisin and construction indus-

* The substance of this article has been drawn from the Stanford Space Systems Engineering Final Report for 1966, *SPINMAP,* which is a satellite system for making long-range weather predictions, School of Engineering, Stanford, California.

tries served as guides. Finally, the study was substantiated by mailing questionnaires to an international sample of companies, agencies, and farmers to obtain the viewpoint of actual users of weather information.

In view of the state of knowledge of the potential value of improved weather forecasts, an estimate of the benefits due to the SPINMAP system is inevitably subjective. However, the conclusions of the following sections are extremely conservative. They concern only those industries listed above, and not the total economy. No attempt has been made to normalize the conclusions to the whole economy, since the benefits based on these few sectors of the economy are striking enough and amply support the ultimate conclusion that benefits from SPINMAP will far exceed the cost.

It should also be stressed that in making these estimates, only *recoverable* costs were listed as benefits. The total cost of the destruction caused by a hurricane can hardly be claimed to be overcome by implementation of SPINMAP, which is designed to predict but not to prevent hurricanes. However, the incremental decrease in damage provided by action taken as a result of improved weather forecasting is considered a benefit as is the increased efficiency of operations resulting from advanced knowledge of the weather.

Participating in the world weather system on the part of the developing nations involves the sociological principles of innovation and change.

First, it is necessary to view the weather forecasting systems in these areas. At present, forecasts, where available, are not sufficiently used. Basically, this is due to a lack of communication and poor education. Any benefit from the forecasts depends on the information the individual receives via radio or newspaper, if available, or from the historical perspectives of experience and folklore. It is up to the individual farmer to assess such information and use it to his advantage.

Into such a pattern practiced for a number of years is introduced information from a previously unknown source. Once reliant on past experience and meager information, a farmer is now asked to accept the new data and to perhaps alter established patterns of planting and harvesting. As with any change, these questions have to be considered: (1) What will this new information replace? Possibly it will be completely new, or it may replace the local "weather consultant." (2) Have the losses caused by weather been so severe that a need to change is felt? (3) Are the possible benefits credible? Can sufficient accuracy be attained so that the proportion of correct forecasts is great enough to convince farmers to alter their activities? (4) To what extent are planting and harvesting isolated activities? Even if farmers were to become convinced of the economic benefits of planting and harvesting at the most favorable climatological times, could they be persuaded to divorce such operations from religious festivals and family tradition? (5) What previous innovations (such as the introduction of farm machinery or fertilizer) have occurred? Have they been readily accepted and what was the rate of innovation?

The sensitivity of agriculture to the weather is easily illustrated by the world's most important crops. Cereal crops are grown in 70 percent of the worlds' harvested areas. They provide more than half of the calorific value of man's food directly and support a large proportion of the other half indirectly in meat, eggs, milk, and so forth.

Cereals, then, are the most important single basic food and crop, and weather is the major risk to the yield level. The connection between weather and yield is emphasized by extreme examples. In 1954, in Missouri, the corn yield was half of the normal crop because of drought. The Soviet Union, usually an exporter of wheat, was forced to import extensively in 1963–1964 owing to poor weather conditions in the newly cultivated virgin lands. To what extent would these misfortunes have been avoidable with improved long-range forecasts?

Studies of the percentage standard deviation of the yield over a 35-year period indicate a value in all growing areas of 0.10 and in some regions above 0.20. Regression analysis indicates that over 90 percent of this deviation can be attributed to weather. Thus three years in ten the harvest cannot be predicted to within 10 percent, and once a century the error is 30 to 60 percent.

A comparison of the yields of wheat in widely separated areas shows that the harvests and therefore weather are not significantly correlated. Hence, weather as it affects crop yields is not global.

An accurate assessment of benefits, of course, is impossible. Even if crop responses are fairly universal, climates and soils are too varied for more than an approximate estimate without intensive study. However, the benefits in the arctic and tropical regions from improved weather forecasts are believed to exceed those of the temperate regions. There are two related reasons for this. First, the forecasts are more fully developed in the temperature zones, and second, farming techniques in the arctic and tropics are more conservative simply because the risks of inclement weather are greater. Replies to the questionnaire and the literature studies indicate that losses up to 50 percent of crop are far more frequent in these extreme geographic regions, substantiating this belief.

These figures indicate that knowledge of the weather itself may be sufficient to better predict food harvest in many areas. This information alone has value in the better planning of food processing and transportation. However, the more direct benefit will come, not from prediction of losses, but from prevention of losses. Better weather information can help a farmer increase his crop yield and/or decrease the labor and materials he uses.

Weather predictions valuable to the farmer fall broadly into three categories. Short-term predictions that inform of impending adverse weather are necessary. The farmer is constantly ready for threats such as frost, and requires about 24 hours warning. Next, predictions which benefit the planning and efficacy of operations may extend over a period of one to two weeks. These forecasts help to specify the optimum timing of plowing, fertilizing, sowing, irrigation, harvesting, and so forth, where information can decrease the risk involved. Finally, long-term trend predictions of over three to four months allow the farmer to assess probable yield from a crop and thereby allow a more effective allocation of people, fertilizer, and land to the most productive crops.

The SPINMAP study concentrated on the use of one- to two-week forecasts. The benefits of one-day forecasts were not considered in detail. Although these benefits are large, most of the United States already has reasonable short-range forecasts and benefits could not be attributed to the improved systems. Benefits associated with predictions of three to four months were also ignored since it is not sure that the advanced systems will make such predictions possible.

Several important world crops were studied in some detail. These included corn and coarse grains, wheat, hay, coffee, and sugar. Though details varied, the general approach was to identify the various steps involved in growing the crop. The effect of weather on each step was then determined, and the action farmers would take with and without long-range predictions was estimated. With weather and crop-loss statistics these estimates were translated into average expected benefits. The results show that the major benefits will come from quite unsophisticated use of weather information. Plowing before heavy rains, reaping a hay crop before several days of sunshine, and eliminating unneeded irrigations are common examples. The benefits estimated in this manner ran from 7 to 15 percent of crop value in the major crops and even higher in some exceptional examples. Over 200 questionnaires returned by farmers show that they feel benefits will be even greater than this.

To simplify the estimates and to introduce a conservative bias into the figures, the SPINMAP group assumed a 5 percent minimum benefit for the crops they considered. Use of the 5 percent estimate for world agricultural benefits was felt to be reasonable since well over 75 percent of the world production of the crops is in the developed countries or in highly organized sectors of underdeveloped countries. Even on subsistence farms the savings should exceed this if the communications systems proposed in SPINMAP are implemented. Most protective action is of an unsophisticated nature apparent to any farmer. Furthermore benefits associated with one-day forecasts should be added for much of the world since even short-range forecasts are presently inadequate in most developing countries.

To give an indication of the magnitude of figures involved, the value of several world crops are shown in Table 1. These figures are from

TABLE 1
DOLLAR VALUE OF WORLD CROPS

Crop	Quantity Produced (million metric tons)	Price (U.S. $/ton)	$ Value (billions)
Cereals			
Corn and coarse grains	450	$ 50	$22
Rice	250	50	12.5
Wheat	250	60	15
Coffee	60 (million bags)	33/bag	2
Grapes	50	100	5
Sugar	30	100	3
Bananas	22	100	2.2
Tomatoes	17	30	0.5
Cotton	11	6	0.06

the *FAO Yearbook* (1964). Table 2 contains the minimum world savings based on the conservative 5 percent estimate. These annual savings of more than $4 billion exceed system cost by well over a factor of ten. Even agricultural benefits to the United States alone, over $1 billion a year, far exceed the system cost. Although study methods vary, estimates done since SPINMAP also come to savings of the same order of magnitude.

Although all human activities are affected in some manner by the weather, there is probably none more vitally affected than aviation. Commercial airlines have become one of the main means of travel in our time and at present represent an industry with an annual gross income in excess of $10 billion.

Thousands of airliners fly daily, and each trip is preplanned to afford the optimum in safety, passenger comfort, on-time arrival, and minimum operating cost. It is in the area of minimum operating cost that the value of a weather forecasting system must be evaluated to determine its true economic benefit to this industry.

TABLE 2
SUMMARY OF BENEFITS TO AGRICULTURE

Crop	$ Value of World Crop × 5 percent (billions)	$ Benefit (millions)
Cereals		
Corn	$22	$1100
Wheat	15	750
Rice	12.5	625
Hay and other feed crops	25	1250
Coffee	2	100
Grapes	5	250
Sugar	3	150
Bananas	2.2	110
Tomatoes	0.5	25
Cotton	0.06	3
Total Dollar Benefit to Agriculture		$4363

At present within the United States, most commercial carriers make extensive use of weather and high-altitude wind forecast, in conjunction with digital computers to compute and select on-time, minimum cost routes. American Airlines reported that the success of their initial project in this effort, begun in 1962, was so astounding that jet aircraft flight planning, on flights greater than 250 miles, was assigned to the computer. From mid-summer of 1962 until June 1963 about 210 flight segments were accomplished. The preliminary analysis of the results suggested annual savings, on fuel and other direct-cost items, of the order of several million dollars. Savings in direct operating costs ranged from a few dollars to over $200 per trip.

The basis for the computer's ability to provide the optimum on-time, minimum-cost routes is the meteorologist who creates the prognostic charts, which form an integral part of the computer program. If in fact he can provide more accurate weather and wind information, then the computer can more accurately select the optimum route and altitude. It has been stated that wind and temperature forecast within the United States are sufficiently accurate that they do not justify further refinement. However, in those sections of the globe where there is at present a minimum of weather and upper-air data, such as the great expanse of the Pacific Ocean, considerable savings could be affected by the commercial airlines through similar programs should the weather and wind data be available. It has been shown that a 10 percent savings in operation cost can be effected when route and altitude selection is based upon forecast data rather than climatology. Certainly one would then expect to realize a major portion of the 10 percent savings in operating cost for those areas of the globe where there is little or no weather and upper-air data. Flights within most of this area are particularly suited to the optimum route and altitude section techniques since most are over-water, intercontinental flights.

The proposed system of weather forecasting will also be useful to shipping companies since it provides the opportunity to chart from sea-state prognostications the most economical ship route before sailing, and striking reductions in operating costs can accrue. Information on storm and cyclone paths, including wind strength and direction together with tides and the residual wave system or swell will be collected. Forecasts of the expected sea state can be made from these data 10 to 14 days ahead with some accuracy and can be updated. Estimates of visibility, including precipitation and fog, can also be derived from the network sources.

On inquiry for a recommended route, an ocean-going ship will need to provide or file special operating characteristics of the ship and cargo. When it is total cost rather than time that a shipowner wishes to minimize, the characteristic variation of fuel usage with speed and wave height are required. The influence of wind and displacement must also be included. The maximum (if uneconomical) speed may allow the possibility of outrunning a storm to be considered; a fragile cargo may limit the maximum mean waves the captain wishes to experience. Generally the operating and fixed charges will be needed to perform the route calculation. The specified route can take account of iceberg warnings, and for passenger liners sunshine routes can be provided.

Computers can calculate the wind and sea state from given meteorological factors for a point at sea. Satellite measurements of sea state should enhance experimentation and by experience should extend accuracy to the full extent of reasonable prebaratics. Variations in sea temperature may also influence the extended accuracy of wind and wave state.

The system of ship routine proposed will also bring benefits in lessening cargo damage, so that it may be possible to reduce cargo insurance rates, currently 1 to 5 percent of cargo value. It is probable that insurance companies will favor the shipping lines that support the route inquiry service.

A further benefit is also expected from the possibility of saving some of the 100 ships (approximately) that are lost at sea each year, but monetary returns are hard to estimate.

A ship routing service constitutes one type of service for which the user might be expected to pay for information required. No estimates of the potential benefits resulting from these calculations appear to have been made. But we anticipate reductions in transit time and cost, together with reduced danger and insurance risks.

The work done for the SPINMAP project involved a more comprehensive and more detailed economic analysis than has been included here. At each step the most conservative estimate of benefits was used, so although it is impossible to arrive at an accurate figure, it is quite possible to arrive at a lower boundary for the economic benefits which will result from the global weather network. The results of the analysis indicate that there would be an overall savings of at least 5 percent, which in the light of the figures of Tables 1 and 2 amounts to a sum far greater than the cost of the project.

The principal conclusions of the SPINMAP study are as follows:

(1) The ratio of benefits to cost of the proposed system will exceed 10. This figure has been derived on the basis of a very conservative estimate of realizable dollar benefits. No effect has been made to place a monetary value on political, social, or other intangible factors which will contribute significantly to the benefits derivable from the implementation of SPINMAP.

(2) As a purely commercial investment, that is, using the proposed system to replace today's expensive observation systems, SPINMAP offers the U.S. government a financial return of 5 percent assuming that the United States finances the entire project. This is not the proposed method of implementing the system, but serves simply to demonstrate that SPINMAP is a sound investment.

(3) If the political problems inherent in international cooperation can be overcome, all nations of the world stand to benefit from the proposed system. SPINMAP provides a mechanism for technical aid and education to the emerging countries and thus constitutes a major opportunity for world cooperation.

components of the satellite system

HAROLD ROSEN

*Harold Rosen received the B.E. degree from Tulane University and
the Ph.D. degree from the California Institute of Technology in electrical
engineering. He is currently Assistant Manager of the Space Systems Division of the Hughes Aircraft Company, Los Angeles, and is in charge
of satellite systems development. He has received awards from the National
Space Club, the American Academy of Achievement, and the National
Academy of Television Arts and Sciences for his contributions to the
development of communications satellites. Dr. Rosen has also been active
in the fields of airborne radar and antiaircraft missile development.*

Abstract—*Part 4, Components of the Satellite System, begins with Dr.
Rosen's discussion of the development, launch procedure, and operation
of two representative satellite systems; the Syncom and Early Bird. The
articles in this part discuss the various components common to many
types of satellite systems and provide an introduction to Part 5, which
deals with systems devoted primarily to the observation of weather
phenomena.*

Communication Satellites

SYNCOM and Early Bird are spin-stabilized synchronous communication satellites which are now in operation over the equator. A synchronous satellite is one that orbits at an altitude of 22,000 miles with an orbital period equal to that of the earth's rotation. In addition to being at this altitude, the orbit is equatorial and circular so that the satellite appears fixed when viewed from the earth. It does not rise or set, but sits motionless in the sky. Such an orbit has many advantages. A single satellite can be used to provide continuous communication links over 40 percent of the earth's surface; only three of them are necessary for complete earth coverage. In addition, the earth terminals used with these systems can be fixed and only one is necessary at a site instead of several as would be required with other systems.

These advantages, and there are many others, of the synchronous satellite have long been recognized, but for a long time it was felt that the problems associated with injecting a satellite into such an orbit and maintaining it there would be too difficult to overcome with current techniques. So the synchronous satellite was relegated to the future, as an ultimate system to be employed at a distant, unspecified time. In the meantime, designers decided to work on systems that presumably could be installed earlier: low- and medium-altitude satellite systems requiring, not one or two or three, but an orbiting fleet of satellites in order to establish continuous communication. They studied systems at low altitudes, medium altitudes, at all inclinations ranging from equatorial to polar, and at all eccentricities. But in the end it was clear that the synchronous satellite, using the simple spin-stabilizer approach which we shall discuss below, was not only practical, but was in many respects easier to use than the other systems. In fact, the ultimate system has become the initial system, both for the Department of Defense, which is using the SYNCOM, and for the Comstat Corporation and the International Consortium of which it is a member, which are using the Early Bird satellite for their initial communications systems. The following figures show how the spin-stabilized, synchronous satellites work.

COMSAT, probably the world's first communications satellite, was built with private funds six years ago (Figure 1). It is designed to spin around a vertical axis defined by the transmitting antenna and has its solar cells placed continuously around its circumference. On the top of the satellite is one of the control jets, along the bottom edge is the V-beam solar sensor used to determine the spin phase and the angle between the sun line and the spin axis. The small black bulge visible at the bottom is the nozzle of the retrorocket used to establish a circular orbit once the satellite reaches altitude.

Figure 1   View of COMSAT satellite.

The COMSAT satellite never flew, but it grew into the SYNCOM program which was sponsored jointly by NASA and the Department of Defense. Figure 2 shows the SYNCOM satellite, which looks very much like the older COMSAT. SYNCOM has on the spin axis a microwave antenna, the tip of which is used for receiving and the base for transmitting. Like COMSAT it uses a V-beam solar sensor and uses solar cells for power. The four whip antennas at the base of the satellite are part of a separate telemetry-command system that operates at 136 MHz and 148 MHz, respectively, in its two phases.

Figure 3 shows the internal configuration of SYNCOM including the orientation jets, the sun sensors, the hydrogen peroxide fuel tanks, and the various communications packages.

Figure 4 shows the principle of the basic two-jet control system used to provide all the orbit and orientation controls necessary for the syn-

Figure 2   SYNCOM.

chronous orbit. Figure 4(a), the orientation control, is used in two modes. In the pulse mode, it is actuated over a small sector of about 60° of the spin cycle in each of several succeeding revolutions. Because it is aimed parallel to the spin axis but offset from it, it produces a torque which causes a slight precession of the satellite with each pulse. In about 100 pulses it can change the orientation of the spin axis on the order

Figure 3   Labeled internal view of SYNCOM.

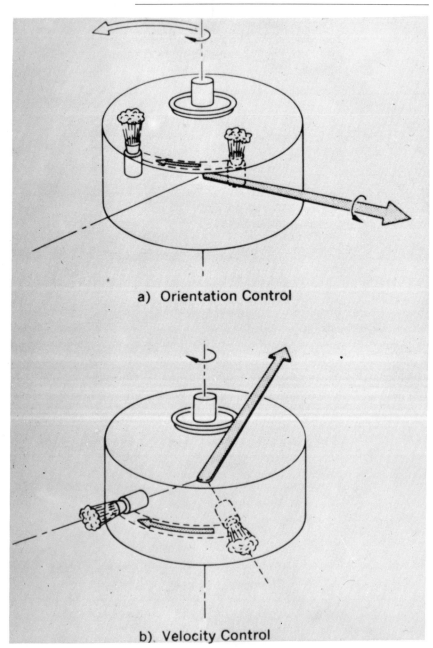

a) **Orientation Control**

b). **Velocity Control**

Figure 4   Schematic of SYNCOM orientation and velocity control.

of 90° if that is desired. This jet is required, not only for establishing the initial orientation, but also for countering the subtle effects of solar radiation pressure which tend to offset the satellite a few degrees per year. In addition to the pulse mode, this jet can be used in a continuous firing mode, in which it is left on for a complete spin cycle. In this case, the torques average out and the net result is just a force parallel to the spin axis. Since in the normal position of this satellite the spin axis is perpendicular to the orbital plane, leaving this jet on for a certain number of complete cycles gives forces which can control the inclination of the orbit. Thus, this one jet is the inclination as well as orientation control.

Figure 4(b) shows the velocity-control jet which fires only in the pulse mode. By choosing the sector of the spin cycle in which to fire the jet, the direction of the force in the plane of the orbit can be chosen. In this way it can change the period of the orbit and keep the eccentricity equal to zero for the circular synchronous orbit. Thus these two jets are able to stabilize the spacecraft entirely. This is one of the basic simplifications achievable in spin configurations as opposed to a three- or four-axis attitude control configuration.

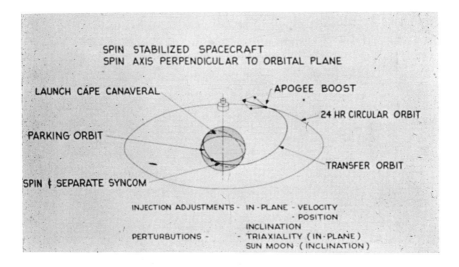

Figure 5   Injection sequence to SYNCOM synchronous orbit.

Figure 5 shows the series of operations required to achieve the synchronous orbit. Starting with the launch at Cape Kennedy, the Delta rocket injects the satellite booster package into a low-altitude parking orbit. As the satellite passes over the equator, the Delta third stage ignites and places SYNCOM in its transfer orbit out to a radius of 22,000 miles. Since the perigee of this highly elliptical orbit is at the equator, the apogee will also be approximately at the equator as required for the SYNCOM synchronous orbits.

The spin-stabilized Delta third stage imparts the required spin to the satellite and then separates from it early in the transfer orbit. At the apogee of the transfer orbit the SYNCOM apogee rocket imparts a velocity increment of about 5000 feet per second, which causes the resulting velocity to be in the equatorial plane and places the satellite in a circular orbit synchronous with the earth.

When the retro maneuver is complete, the orientation control jets are used in the pulse mode [Figure 4(a)] to reorient the spin axis perpendicular to the plane of the orbit. Then a combination of the orientation and velocity-control jets advances or retards the satellite in its orbit relative to the earth to adjust its period and to focus its beam at the desired longitude on the earth's surface. This self-contained control system must then continually correct for the perturbations due to the sun-moon attraction out of the orbital plane and for the east-west drift due to the eccentricity of the earth's equator.

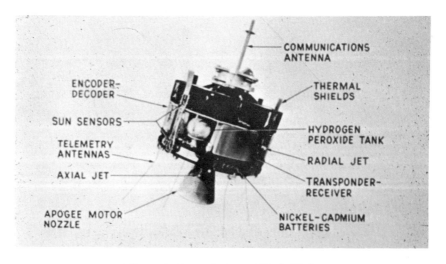

Figure 6   Internal view of Early Bird.

Figure 6 is a view of the internal structure of the Early Bird satellite which looks very much like the COMSAT and SYNCOM satellites. It is larger than SYNCOM but can still be launched by the Thor Delta rocket. Early Bird generates about 60 percent more dc power than SYNCOM with $1\frac{1}{2}$ times as much solar panel area, and its traveling wave tube is somewhat more efficient than SYNCOM's, developing about 6 W of power. The antenna has more directivity with 9 dB, or about twice the gain of SYNCOM. However, it does not provide complete north-south coverage so that the beam is aimed at the North Atlantic, where the present traffic requirements exist.

Figure 7 shows the utilization of the frequency spectrum by the Early Bird satellite. There are two bands, one for transmission between Europe and the United States and another for the reverse path, between the United States and Europe, each of which has about 25-MHz bandwidth. The two up-links are in the 6000-MHz band; the two down-links are in the 4000-MHz band and they are separated by about 100 MHz.

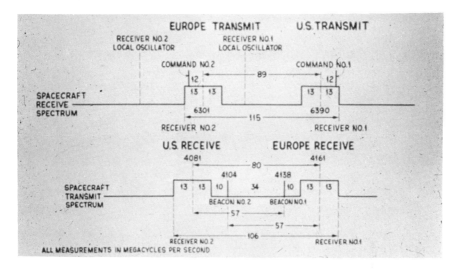

Figure 7   Early Bird communication RF spectra—Europe–U.S. link.

Figure 8 shows the antenna contours of equal density on the ground and shows how the area between the northeast United States and Western Europe is favored by the particular antenna patterns shown.

Figure 9 shows the location of SYNCOM II, SYNCOM III, and Early Bird. All three satellites are in synchronous equatorial orbits with SYNCOM II over the Indian Ocean, SYNCOM III over the International Dateline, and Early Bird over the Atlantic Ocean. SYNCOM II was launched before the augmented Delta rocket was available, so there was not quite enough thrust to remove the inclination corresponding to the latitude of the launch site. The solid curve at the bottom of the figure shows the location of the maxima of potential of the earth's gravitational field. Because the equator is not round, a satellite initially in a synchronous orbit will start drifting either eastward or westward depending upon which side of the potential hill it is on. There are two stable points, one at the longitude of Mexico and the other above India, and two unstable points of equilibrium at the two peaks in the curve. SYNCOM II is over

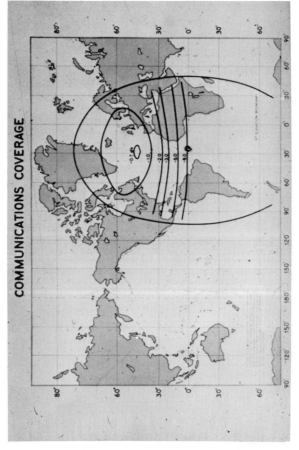

Figure 8　Early Bird communications coverage superimposed on map of the world.

74

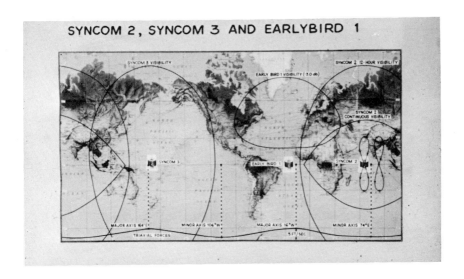

Figure 9   Positions of SYNCOM II, III, and Early Bird.

the Indian Ocean at a longitude corresponding to a stable equilibrium, and even though the relatively primitive control system has long since run out of fuel, it will forever remain in this general area. Early Bird is not far from an unstable equilibrium point. However, it has a better control system and we expect to be able to control it for many years.

These three satellites have by now accumulated almost five years of total operating time with no major failure. In fact, there has been only one electronic failure in SYNCOM II in one of the two redundant telemetry systems. At the time this program started, the longest satellite lifetime had been about one month. The favorable environment that exists at the synchronous altitude, and the equable temperature due to spinning, probably accounts for this longevity. The satellite is almost always in the sunlight (the eclipses are very infrequent—less than 1 percent of the total time is not in the sunlight), and the spin stabilization promotes equal temperature environments of about 70°F throughout the satellite, so that there is no hot or cold side as there are in other satellites.

Figure 10 is the spin-stabilized version of the ATS (Application Technology Satellite). This program has had a long history. It grew out of a program called the advance SYNCOM which started in 1961, soon after the basic SYNCOM program had started, but the advanced SYNCOM had to be modified when the COMSAT Corporation was

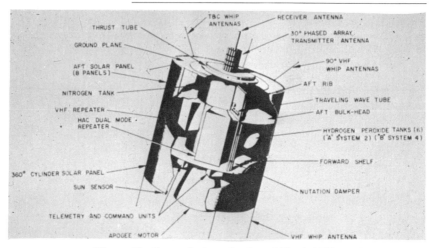

THRUST TUBE
GROUND PLANE
AFT SOLAR PANEL (8 PANELS)
NITROGEN TANK
VHF REPEATER
HAC DUAL MODE REPEATER
360° CYLINDER SOLAR PANEL
SUN SENSOR
TELEMETRY AND COMMAND UNITS
APOGEE MOTOR

T&C WHIP ANTENNAS
RECEIVER ANTENNA
30° PHASED ARRAY TRANSMITTER ANTENNA
90° VHF WHIP ANTENNAS
AFT RIB
TRAVELING WAVE TUBE
AFT BULK-HEAD
HYDROGEN PEROXIDE TANKS (6) ('A' SYSTEM 2) ('B' SYSTEM 4)
FORWARD SHELF
NUTATION DAMPER
VHF WHIP ANTENNA

Figure 10   Internal view of spin-stabilized ATS.

formed and the government, NASA in particular, was no longer permitted to pursue the development of communications satellites. So the advanced SYNCOM became a general-purpose applications technology satellite which carried many experiments, only a few of which were directly related to communications. The structure of the ATS is necessarily more flexible to accept the new experiments, but basically it is the same as that of SYNCOM and Early Bird. One new experiment is the meteorological camera which employs the spin of the satellite for its horizontal scan.

The meteorological camera, or the SSCC (Spin-Scan Cloud Camera) which will be used in the ATS satellite, is a tube with a 5-in. diameter reflector. It faces radially outward from the satellite perpendicular to the spin axis. Hence the earth sweeps across its field of view once each revolution. The camera is mounted on a platform attached to a small stepping motor to move the camera's field of view from north to south over the earth's surface. Both the resolution of the camera and the step-width per revolution of the motor are one-tenth of a milliradian. Even though its mass is far less than that of the satellite, the camera is very carefully placed so that its motion will cause no perturbation in the satellite's attitude. The video information from each scan is fed onto a voltage-controlled oscillator and transmitted to earth via the communications transponder.

At the synchronous altitude of 22,000 mi, a resolution of one-tenth of a milliradian implies a resolving distance of about two miles. This system will be used to determine the velocity of cloud patterns and the wind vectors over the oceans when there are clouds present.

Figure 11 shows the ascent sequence for the ATS spin-stabilized version.

It is identical in almost all respects with that used for the Delta-launched satellite, the one significant change being that the ATS is launched by the Atlas-Agena instead of by the Delta. The Atlas-Agena can lift about 10 times as much payload as the old Delta, or about 5 times as much as the new Delta, but since its upper stage does not spin, the spin has to be imparted to the satellite by a self-contained spin-up system which uses a nitrogen bottle. This is the only major difference between the ascent sequence for the ATS and the SYNCOM and Early Bird.

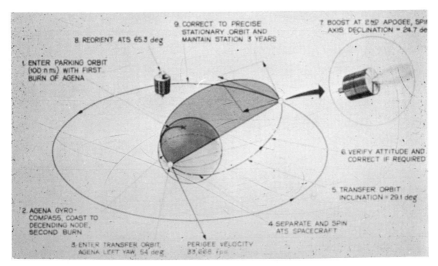

Figure 11   Ascent sequence for ATS.

The ATS satellites contain a dual mode communication transponder: the frequency translation mode (which has gone into the Early Bird), which is a 25-MHz bandwidth channel; and the multiple access mode (which was an attempt to provide a means for handling simultaneously, and with efficiency, signals arriving from a large number of independent earth stations). In fact this mode can handle signals from 1200 different earth terminals. The format of the signals is single sideband for the ground-to-air link and the satellite processes them in such a way that they all are combined into one carrier frequency and return to earth as one wideband-frequency-modulated wave. In this manner, there are no significant intermodulation products to be concerned with on the down-link which was considered to be the most difficult when the system was designed.

The earlier SYNCOM antenna had no directivity about the spin axis. It thus had a fairly low gain in the direction of the earth, but it was a simple reliable component which worked well on the earlier communication satellites. One way of improving the gain is to use not one but several antennas, all similar to the one used on SYNCOM, arranged in parallel around the spin axis. If the phase of the signal fed to each antenna is adjusted to exactly compensate for the different distances of each antenna from earth, then the resultant signal will form a beam in the direction of earth. Such a system of simple parallel antennas provides a suitable azimuth gain for the ATS satellites.

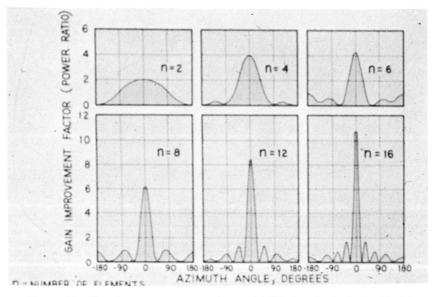

Figure 12　Phased array antenna patterns (ATS) for 2, 4, 6, 8, 12, 16 elements.

Figure 12 shows the various antenna patterns around the spin axis which result from antennas having 2, 4, 6, 8, 12, or 16 elements. In the case of 16 elements, which is the number used in the advanced SYNCOM, the gain improvement is over 10 times. In the case of 8 elements, which is the number of elements used in the VHF repeater in the ATS satellite, the gain improvement is over 6 times. Figure 13 shows the construction of the phase array antenna of the ATS satellite. The 16 elements are driven by the 8 phase shifters, of which you can see the front 4. Each phase shifter has two outputs which are directed by the strip line circuit to diametrically opposite antenna elements. This is an all-electronic system which generates a beam stabilized in space along the satellite-to-earth line. Traditionally, one of the problems of a spinning

Figure 13 Phased array antenna.

satellite was the difficulty of obtaining a directional antenna pattern. With the ATS system the beam is pointed toward earth and remains fixed as the satellite spins under it.

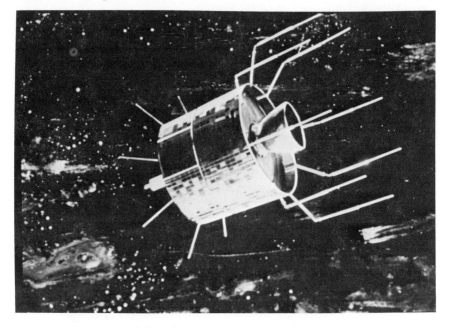

Figure 14   Artist's view of ATS satellite.

Figure 14 is an artist's concept of the first ATS satellite. On the left is the high-gain microwave antenna system and on the right are the eight elements of the VHF repeater. This has an effective radiated power of about 300 W, which is a big improvement in VHF communications and permits voice channels to be transmitted to and from airplanes.

One of the descendants of the ATS program has been the Educational Television Satellite shown in Figure 15. The objective of the ETS program is direct communication between the satellite and an individual isolated school. Such a program has become feasible because of the larger payloads made possible by the Atlas-Agena booster.

If one goes back and starts over with a payload optimized for the Atlas-Agena and devotes all that payload to communications, one finds that the effective radiated power that could be produced goes up by several orders of magnitude. It would now be possible to transmit a television picture directly from the satellite to a user who was equipped, not with an 85-ft dish, but only with a small antenna that any individual could easily afford. To do this one must simply increase the power from the

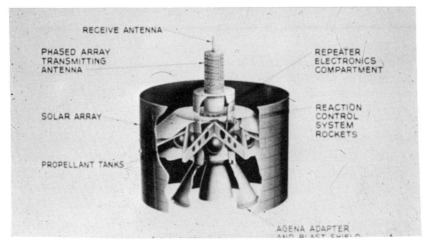

RECEIVE ANTENNA

PHASED ARRAY
TRANSMITTING
ANTENNA

REPEATER
ELECTRONICS
COMPARTMENT

SOLAR ARRAY

REACTION
CONTROL
SYSTEM
ROCKETS

PROPELLANT TANKS

AGENA ADAPTER

Figure 15   Internal view of ETS satellite.

satellite; no other new techniques are required. Using the shroud from the OAO (Orbiting Astronomical Observatory) satellite, which was developed for the Atlas-Agena, a payload diameter of 9 ft is permitted. Using that shroud, one could use solar-cell arrays which can develop about 600 W of solar power which, allowing for degradation over many years of orbit, could power 16 traveling wave tubes generating a total of 100 W of power. Furthermore, with the adaptations of the phased array transmitting antenna discussed above, one could form much narrower beams with gains up 1000 or even 10,000 times. Thus a beam with an effective radiated power on the order of 100 kW could be formed with a satellite launched by the Atlas-Agena.

With this ETS system video programs, particularly educational video programs, could be transmitted to schools in many areas of the world where no communications systems now exist. The ground station would need only a small source of power, a 6-ft receiving dish on the roof, and a radio monitor in the classroom to put it in touch with every educational center in the world.

An ETS satellite with an antenna gain of 30 dB would have a beam width of 5° and a power output of approximately 100 kW. At an altitude of 22,000 mi, a beam of 5° would cover about 30° of longitude or latitude on the earth. One satellite could thus communicate with all of a country the size of India or Mexico, two satellites with a country the size of the United States. This application of our communication and space technology could very well be the most significant result of our entire space development effort.

EDWIN MARRIOTT

Edwin Marriott received an M.S. degree in aeronautical engineering from Stanford in 1950 and has since worked in several departments of the Hughes Aircraft Company. He was Manager of the Aerodynamics Department and, subsequently, Engineering Mechanics and Preliminary Design Departments. In 1964 he was appointed to his present dual position of Assistant Manager of the Communication Satellite Systems Laboratory and Manager of the Applications Technology Satellite program (ATS).

Abstract—*The ATS program grew out of the earlier SYNCOM satellites discussed above by Dr. Rosen, and as its name implies, serves as an orbiting platform for a variety of experiments, among which is the spin-scan cloud camera described in Part 5 by Dr. Suomi.*
Mr. Marriott begins with a general discussion of the ATS program and then concentrates on the gradient-stabilized vehicles and the problems associated with this particular form of stabilization.

Gravity-Gradient Satellites of the ATS Program

THE broad objectives of the ATS program are to achieve a long-life spacecraft capable of carrying a wide variety of experimental payloads, to explore synchronous and medium-altitude orbits, to carry some 20 major experiments, and to explore gravity-gradient stabilization at high altitudes. Hughes Aircraft Company has been under contract with the Goddard Space Flight Center since May 1964 on this program.

The program consists of five satellites in three different configurations. Two are spinning satellites, the ATS B and C, which are quite similar to the old SNYCOM family, although larger, being 5 ft in diameter and about 6 ft long. There are two synchronous-altitude gravity-gradient-stabilized vehicles and finally, a medium-altitude gravity-gradient-stabilized vehicle. The main purpose of the medium-altitude system, which is at 6000 mi in an inclined orbit of about 28°, is to explore the characteristics of the gravity-gradient stabilization at high altitude in enough depth to fully understand the stabilization field and the disturbing forces. Most of the gravity-gradient-stabilized vehicles that had flown as of 1967 have been at low altitudes, have been for feasibility studies to test the principles of gravity-gradient stabilization, and have carried very little instrumentation. The ATS satellites will contain sufficient instrumentation to get good data about the stabilization field and disturbing forces. The first spin-stabilized satellite was launched at the end of 1966 with the four subsequent shots following at six-month intervals.

Table 1 lists several of the 20 experiments flown on the five ATS satellites. Dr. Rosen elsewhere in this chapter describes in some detail the electronically despun antenna used to increase the antenna gain on the first of the spin-stabilized satellites. The second spinning ATS employs an interesting mechanically despun antenna.

Some of the meteorological experiments require a very accurate stabilization of the spinning platform. A nutation sensor, which is simply a

TABLE 1
APPLICATIONS TECHNOLOGY SATELLITE
(ATS) EXPERIMENTS

Albedo (two)
EME (environmental measurements) three
Gravity-gradient stabilization and instrumentation (three)
Ion engine (four)
Mechanically despun antenna
Meteorological (two)
Nutation sensor
Spin-scan cloud camera
VHF repeater
Other experiments (not assigned now)

very sensitive accelerometer, measures the small nutations (on the order of a hundredth of a degree) of the platform. The spin-scan cloud camera which Dr. Suomi describes in Part 5 is used in the ATS satellite for gathering weather data. The VHF repeater or transponder is used to communicate with aircraft and to transmit weather information via the satellite between ground stations.

Figure 1 summarizes the three methods of spacecraft stabilization. The Nimbus satellite, for example, uses an active control system consisting of sensors, reaction wheels, and gas jets to hold a given attitude about all three axes. Passive stabilization systems include the gravity-gradient system which will be covered in more detail below, a randomly oriented uncontrolled satellite and a spin-stabilized system such as the SYNCOMS. The advantage of the passive system, of course, is that it is less complex and hence potentially more reliable. Each method of stabilization has different characteristics which must be weighed carefully in selecting a control system for any particular program. The spin-stabilized system will remain in a fixed attitude with respect to its orbit for a relatively long period of time, and the gravity-gradient-stabilized satellite has the peculiar feature of providing passively an earth-oriented axis which eases the problem of pointing cameras and antennas. Between the two extremes are the semipassive systems.

The objective of the ATS gravity gradient experiment is to investigate and define

(1) The optimum configuration of the satellite and satellite stabilization system.

(2) The most efficient damping technique.

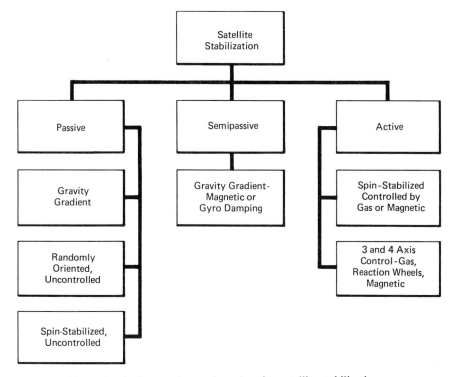

Figure 1   Active, passive, and semipassive satellite stabilization.

(3) A reliable boom extension system and capture procedure during which the long stabilization booms are unwound and the space-craft stabilizes in this gravity gradient mode.

(4) The stabilization tolerances, particularly at high altitudes for gravity gradient stabilization.

During the launch when the satellite is enclosed in the rocket shroud the long stabilization booms obviously cannot be fully extended. Once the satellite is in orbit in some random orientation the booms must be extended in a manner that will not allow forces to build up in the thin, metallic structures which might break them. The term "extension" denotes this process of deploying the stabilization booms. After the booms are deployed, the stabilization forces begin to affect the randomly rotating spacecraft and will eventually "capture" the spacecraft, thus producing the earth-oriented axis.

Figure 2 is a cutaway model of one of the Synchronous Altitude Grav-ity-Gradient satellites. At the high altitudes of either the medium- or the

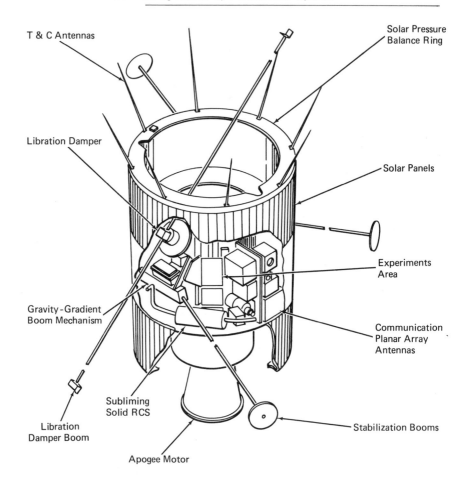

Figure 2   Cutaway view of synchronous ATS.

synchronous-altitude satellites the sun shines on one side of the spacecraft for long periods of time. Hence in the nonspinning gravity-gradient-stabilized models the problems of high solar-cell temperatures (which implies a sizable loss in efficiency) on the sun side and large thermal gradients across the spacecraft become quite severe. The Nimbus satellite solves this problem with external solar panels, but at the expense of added control complexity. In the ATS, as shown in Figure 2, the central experimental portion of the satellite contains only thermal control surfaces, and the hollow end sections contain the solar cells, which can now radiate across the hollow interior section to reduce the thermal gradients. This radiation coupling allows a reasonable efficiency for power generation. It also reduces the thermal gradients across the panels from $+250°$, $-250°$

to something like $+120°$, $-80°$. Without such coupling the problem of finding adhesives adequate to hold solar cells on to the panels over the entire temperature range becomes acute.

Figure 2 is slightly out of perspective in that the satellite is 5 ft in diameter and 6 ft long and the stabilization booms are about 130 ft long. The direct evolution of the ATS from the earlier SYNCOM and Early Bird is clear from the overall appearance of the spacecraft.

Figure 3 shows the mission profile for the launch of the medium-altitude

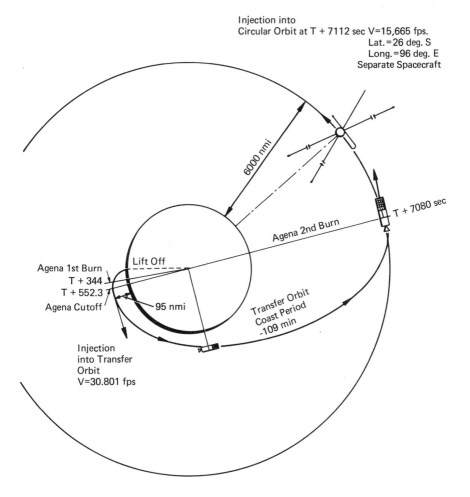

Figure 3   Mission profile for medium altitude ATS.

gravity-gradient satellite. The first burn of the Atlas-Agena puts it directly into a transfer orbit which takes it out to 6000 mi where the second burn of the Agena places the spacecraft in a circular orbit. This is very attractive because it means that the satellite itself needs no apogee kick engine so that for the first high-altitude gravity-gradient system we merely eject the satellite from the rocket and extend the booms. It makes the satellite itself as simple as possible. On the upper right is a picture of the satellite itself in orbit with the long booms in an X configuration. The X-boom plane and the damping boom itself are both inclined to the orbital plane such that the longitudinal axis of the cylindrical satellite is nominally normal to the orbital plane.

Figure 4 shows the added complexity required to go to a synchronous altitude when limited by the booster. Given a big enough booster, one could repeat the procedure outlined in Figure 3; have the rocket enter the transfer orbit directly, go up to synchronous altitudes, circularize the orbit, separate the satellite from the booster, and extend the booms. Since such rockets are prohibitively expensive, the ATS uses a more complex maneuver with the Atlas-Agena booster.

The Atlas places the spacecraft into a 100-nmi parking orbit. The Agena's second burn is just sufficient to kick the spacecraft into an elliptical transfer orbit. The ATS separates from the Agena, and is spun up for the $5\frac{1}{4}$-hour journey to the synchronous altitude. At this point the apogee engine (bottom of Figure 2) fires to put the satellite in a circular equatorial orbit. (This also involves removing the inclination of the transfer ellipse). Small hydrogen peroxide rockets then reorient the satellite, after which it is despun. The gravity-gradient booms then unfurl, and the ATS is in its mission orbit. The procedure is clearly more complex than that for the medium-altitude orbit, which in part explains why the latter was chosen for the first ATS gravity-gradient mission.

Table 2 contains a general description of the synchronous gravity-gradient satellite which is about 1650 lb at launch for the synchronous system and down to about 710 lb in orbit after burning and ejection of the solid-fuel apogee engine. There is a 5-lb thrust hydrogen peroxide control system to get the satellite initially positioned in orbit and some very interesting subliming solid engines (discussed below) which provide station keeping capability once the satellite is correctly positioned in orbit. The solar cells provide about 135 W at 28 V, and several batteries supply peak power demands. The telemetry system has four 2-W transmitters and two separate encoders, and the command system has two receivers and two dual-mode decoders. The communication system has two triple-mode repeaters which include a frequency translation mode, a multiple access

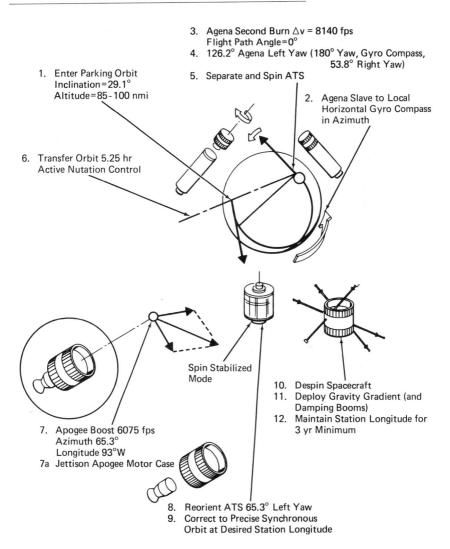

3. Agena Second Burn Δv = 8140 fps
   Flight Path Angle=0°
4. 126.2° Agena Left Yaw (180° Yaw, Gyro Compass,
   53.8° Right Yaw)
5. Separate and Spin ATS

1. Enter Parking Orbit
   Inclination=29.1°
   Altitude=85-100 nmi

2. Agena Slave to Local
   Horizontal Gyro Compass
   in Azimuth

6. Transfer Orbit 5.25 hr
   Active Nutation Control

Spin Stabilized
Mode

10. Despin Spacecraft
11. Deploy Gravity Gradient (and
    Damping Booms)
12. Maintain Station Longitude for
    3 yr Minimum

7. Apogee Boost 6075 fps
   Azimuth 65.3°
   Longitude 93°W
7a Jettison Apogee Motor Case

8. Reorient ATS 65.3° Left Yaw
9. Correct to Precise Synchronous
   Orbit at Desired Station Longitude

Figure 4   Mission profile for synchronous altitude ATS.

mode for communicating with the satellite from stations all over the world, and a wide-band data mode. The frequencies for the up-and-down links are approximately 6000 MHz and 4000 MHz, respectively. There is also an 18-dB planar array antenna.

The basic principle of gravity-gradient stabilization is simple and should be examined for a better understanding of the ATS gravity-gradient satel-

TABLE 2
DESCRIPTION OF SYNCHRONOUS GRAVITY-GRADIENT ATS

56 in. diameter; 72 in. length
1625 lb at launch; 670 lb in orbit
Apogee motor:
    Scaled JPL starfinder
Control systems:
    5 lb H202
    Subliming solid (2)
    (inversion, E-W stationkeeping)
Electrical power:
    Solar panel array (22,080 cells)
    Net power: 135 W
    Rated voltage: 28.3 V
    Batteries (two 22-cell NICAD)
Telemetry:
    Four 2-W transmitters
    Two encoders
Command:
    Two receivers (150 MHz)
    Two dual-mode decoders (255 commands)
Communications:
    Two triple-mode repeaters
    Four 4-W TWTS
    18-dB planar array antenna
    6301, 6212 MHz ground to spacecraft
    4179, 4120 MHz spacecraft to ground
    4136, 4195 MHz beacon frequencies

lite configuration. If a dumbbell floats freely in space about the earth, it will have a tendency to align itself with the local vertical as shown in Figure 5. The reason for such a preferred orientation is that the mass closer to the earth is in a slightly stronger gravitational field than the mass farther away, so there will be a slight torque tending to align the axis of the dumbbell with the earth's radius. Figure 5 gives an idea of the magnitude of such a stabilizing force. Note that since T varies inversely as the cube of altitude, the stabilizing forces decrease rapidly at the higher altitudes. In equilibrium, this "stabilized" altitude provides a vertical reference line by which to point communication antennas, cameras for meteorology, or any other instrument directly at the earth. The presence of an earth-pointing axis is useful. There are several other requirements for a workable stabilization system, one of which is some type of damping.

There is very little damping associated with a simple dumbbell structure.

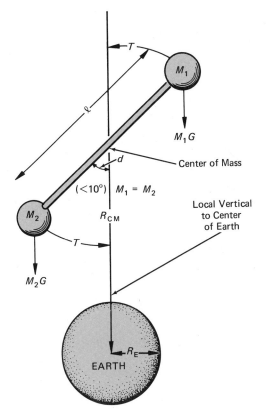

Figure 5   Gravity gradient technology. Note: M_2G is greater than M_1G because of the gravitational field gradient. The result is a torque (T) about the center of mass, tending to stabilize the body along the local vertical

$$\text{net torque, } T = \frac{-\alpha g R_E{}^2 M l^2}{R_{EM}{}^3}$$

Thus, two 250-lb masses separated by 10 ft, orbiting at 5700 nmi, one degree from the local vertical, would have an orienting torque of 6.26×10^{-5} lb-ft/deg., or 850 dyne-cm/deg.

In the frictionless environment of space such systems, when disturbed, oscillate for long periods of time at a frequency determined by the orbit altitude (the amplitude of oscillation increases with altitude as the force of gravity diminishes). In order to stabilize such a structure for use, energy must be extracted from the system so that the oscillations are *damped* and the system eventually approaches the stable earth-oriented configuration.

The stabilization system chosen for the ATS is just a pair of dumbbells placed in an X configuration together with a single damping boom which lies nominally in a horizontal plane. There are two reasons for using the ATS X-boom and damper configuration. First, a single dumbbell right along this axis is in the way of earth-pointing cameras or antennas and an X-boom pattern gives access to the earth axis. The second reason is more fundamental; it gives the third axis of stability. The combination of the damper boom and the X booms gives, not only the earth-pointing axis stabilization, but "roll" stabilization about the earth-pointing axis as well. An earth-pointing axis means that the spacecraft rotates at one revolution per orbit (like the moon with one face always toward the earth). This orbital angular velocity is not large, but it is enough to give some centrifugal force as the satellite rotates about an axis normal to the orbit plane. Splitting the booms will tend to put the X-boom plane in the plane of the orbit because the satellite spins slowly about an axis perpendicular to the orbit. Similarly, the damper boom tends to rotate into the orbit plane, resulting in an equilibrium "yaw" altitude. This gives the added control required for stability around the third axis.

Figure 6 is a picture of one of the dampers aboard the ATS. A magnetic disk is attached to the long booms which rotate about the X-X axis. The whole system is suspended on a torsion wire and as it oscillates, the magnetic discs, going through a local magnetic field, produce hysteresis magnetic damping.

One of the major disturbance torques for a gravity-gradient satellite at synchronous altitudes is caused by solar pressure, which affects the satellite in two ways. The first is that the ATS is not symmetrical when viewed by the sun's radiation. The ideal configuration for minimizing the torques due to solar pressure would be a sphere; however, a sphere is very poor for solar power and for packaging experiments; there just is not very much usable volume. Thus one uses a cylindrical configuration which is perfectly symmetrical from a solar torque standpoint as long as the sun is normal to the cylindrical axis. But, as the satellite orbit moves out of the ecliptic plane, this angle changes by as much as $+23\frac{1}{2}°$. This means that the solar radiation is coming at an angle to the spacecraft which reduces the normal forces and increases the shear forces along the solar panels. This also produces shear and normal forces on the end plane which must be carefully balanced by particular selection of surface properties of the materials used.

The second aspect of solar balancing is the boom bending· problem. With the 123-ft boom on the synchronous satellite, the sun will shine from one side toward these booms, one side becoming hotter than the other, resulting in thermal bending. Part of that bending will be symmetri-

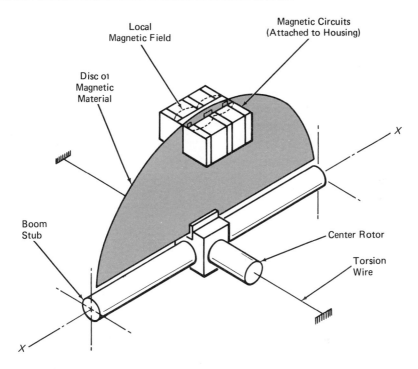

Figure 6   Magnetic damping unit used on gravity gradient ATS.

cal and no solar pressure torques will come from it; however, the boom on one side of the spacecraft will bend to increase the area facing the sun and on the other will bend to reduce the area facing the sun. Thus each boom will have a different force solar pressure which produces a solar-pressure torque. This can be very severe depending upon the amount of bending in the boom. As a matter of fact, this is a factor in determining the length of the booms. One would like them very long to provide a large moment of inertia to provide large stabilizing forces, but as they get very long, the bending increases, so the torques increase, and the disturbances torques become too large.

Another disturbing torque comes from orbit eccentricity. If the orbit is not circular, the angular velocity of the local vertical will vary. The gravity-gradient-stabilized satellite will tend to "track" this local vertical whose angular velocity is alternately increasing and decreasing. The result is an oscillating altitude error about the local vertical in the orbital plane. There are also magnetic and micrometeorite impact effects which disturb the satellite orientation.

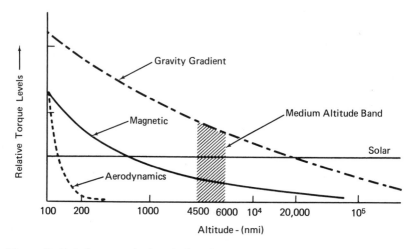

Figure 7   Relative magnitude of disturbance and gravity gradient torques. Note: Actual satellite torque disturbance errors depend on the specific configuration.

Figure 7 shows the relative torque levels as a function of altitude in nautical miles. As the altitude increases, the stabilization forces which are shown by the gravity-gradient-line decrease; however, most of the disturbing forces decrease as well. Aerodynamics are negligible at the ATS altitudes, and magnetic torques may or may not be negligible depending upon the degree of magnetic cleanliness achieved in the spacecraft design. The force that is important is the essentially constant solar pressure torque which ultimately limits the altitudes of the gravity gradient ATS satellites. At the 6000-mi medium altitude the solar pressure is an important influence, at the 22,000-mi synchronous it becomes the dominant force.

Figure 8 shows the basis for the so-called STEM principle, which stands for Storable Tubublar Extenidle Member. The tape used is about $\frac{3}{1000}$ in. thick, about 2 in. wide and is rolled upon the spool to the right of the figure. When the satellite is in orbit, the metallic tape unrolls through the guide rollers and curls up. Because it is initially prestressed, its natural configuration is tubular, so that even if the guides were not there, it would tend to assume its tubular shape; the guides keep it under control. The tape extends for 130 ft, and it is sufficiently flexible that bending of this boom becomes a problem. Most of these booms are beryllium copper, and the first ones had a large temperature difference between the front and back and hence exhibited large thermal bending. To ease that problem, they are now silver-coated so that they absorb less solar energy, have

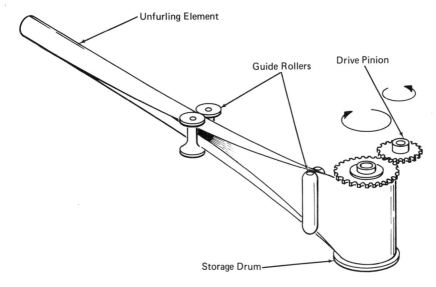

Figure 8   Stem principle for unfurling booms.

a lower temperature difference between front and back, and hence less thermal bending.

One of the interesting features which results from the bending of the booms is that the center of gravity of the satellite is not constant. Figure 9 shows that if the sun is coming in from the top and the satellite is in position A, the X booms will tend to deflect under thermal bending to the dotted position. At position C it is the same, they bend away from the sun. At D and B again they bend away from sun. This means that the center of gravity of the satellite is not constant. On the end of these booms, there are 8-to-10-lb masses in order to get a large moment of inertia for stabilization. When 30 or 40 lb move by several feet, the center of gravity of the satellite must also shift.

This of course would not be any particular problem unless one wished to control the satellite. For the synchronous spacecraft one would like to keep it at the same longitude, and this requires a corrective thrust which imparts a velocity increment up to about 7 ft/sec/yr. This is a very low thrust, but even at such a level, because of the changing center of gravity, significant disturbing torques result. Because the stabilization forces are so low, even very low torques are hard to accept. Ion engines have characteristically worked around 10^{-2} or 10^{-3} lb of thrust. To balance the ATS we have gone two orders of magnitude below this to about 10^{-5} lb of thrust! The ATS uses a subliming solid system which, at the

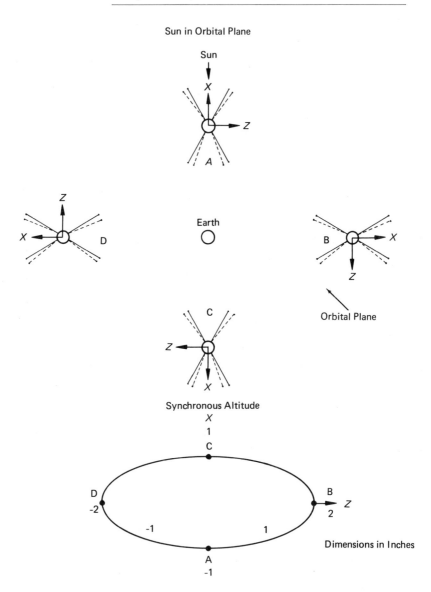

Figure 9   Center of gravity shift from thermal bending of booms.

ambient satellite temperature, has a very low subliming rate. When a thrust is required, heat is provided to the propellant to get it to sublime at a higher rate so that it delivers particles from the sublimation to provide station keeping at 10^{-5} lb of thrust.

These small jets also serve as an inversion system. As I mentioned earlier, when the satellite first enters the gravity-gradient mode, it has residual spin rates from separation dynamics (medium-altitude mission) or despin dynamics (synchronous-altitude mission). The booms open, and since the spacecraft is spinning at some low rate in general, the satellite conceivably could come to rest (capture) either upside down or right side up. The spacecraft is reoriented by the subliming solids, with a pair of the tiny rockets that produce torques in opposite directions. One rocket starts to invert the satellite, gets shut off after about 90° of rotation, the other comes on and slows the rotation down until rotation ceases at 180°, and the inversion is complete. Clearly, time is not of the essence in this process. The thrust for these rockets is 5.4×10^{-4} lb.

The thermal problem on the spinning satellites is almost nonexistent. This is one of the major advantages of spin stabilization. The satellite, because of its relatively high spin rate, operates at about 70° ambient temperature; it varies somewhat with the sun illumination angle but tends to remain quite constant. However, on the synchronous-gravity-gradient satellites which rotate with respect to the sun only once a day, hot spots develop on one side of the spacecraft. Although we were able to reduce the solar-panel gradient across the spacecrafts by allowing radiation coupling, this reduction ($+120, -80°$) is not adequate for most spacecraft equipment. We must therefore use an insulated compartment. We must carefully insulate the inner instrument compartment from the outer panels so that there is a long time constant with respect to one day. The temperature then remains essentially constant within the spacecraft. With this insulation on the ATS there is a maximum temperature difference due to solar effects inside the spacecraft of about 20° between the hot and the cold sides.

With such a large amount of insulation, dissipation of heat from within the spacecraft becomes a problem. If the satellite used a constant amount of power, we could design the leakage of the insulation to maintain a constant internal temperature. However, for a satellite with many different experiments on board, at any given time some of them may be on and some may not; hence the power and the heat levels will vary, depending on the amount of internal activity. Without using power programming, which is an additional operational complication, it is necessary to be able to vary the level of heat transfer to space. Figure 10 shows the ATS active

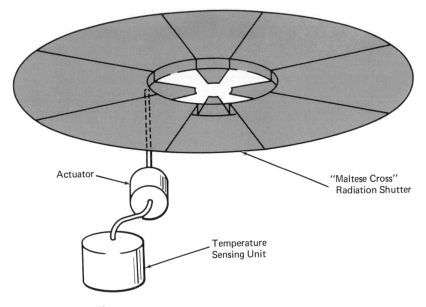

Actuator

"Maltese Cross"
Radiation Shutter

Temperature
Sensing Unit

Figure 10   ATS active thermal control system.

thermal control system, which is simply a method for varying the emissivity of the spacecraft. When the internal temperature rises, it simply increases the amount of heat being emitted. As shown in Figure 10 this is a circular panel with varying levels of emissivity in different areas which can be exposed to or shielded from outer space by rotating the panel. The temperature sensing unit monitors the internal temperature and determines whether the high or low emissivity surfaces of the panel will be exposed for radiating heat into space.

Figure 11 is a graph of the average temperatures in the spacecraft as a function of the power in the satellite. The ATS has a range of about 30 to 100 W. With a passive thermal control system, the temperatures would vary along one of the inclined lines, depending upon the insulation and leakage rate. With an active control system the temperature is almost constant at about 70° to 80°F over this power range of 30 to 100 W.

In summary, the gravity-gradient version of the ATS is a challenging new alternative for high-altitude meteorological and communication satellite systems. The ATS gravity-gradient satellite does suffer some loss of solar-cell efficiency because of the high operating temperature of the panels, it does require an active thermal control system, and there remain uncer-

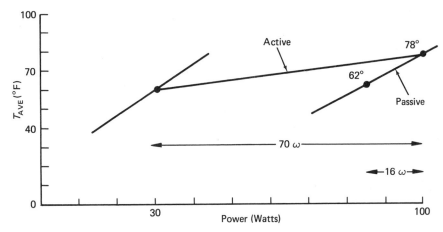

Figure 11   Comparison of internal temperatures versus power level for active and passive thermal control systems.

tainties about the deployment and performance of the stabilization booms. On the other hand, the gravity-gradient technique is simple and reliable, it does alleviate the necessity of a costly active control system, and it does provide a stable earth-pointing axis for cameras or communications antennas. The results of the ATS program will help to determine whether the gravity gradient stabilization technique should be used on future meteorological satellite systems.

KEITH WINSOR

Keith Winsor graduated with honors in 1952 from California Institute of Technology with a B.S. degree, majoring in electrical engineering. He is currently Chief Engineer of the Power System Division of Electro-Optical Systems Inc. (EOS), a subsidiary of Xerox Corporation. He joined EOS in 1960 as Department Manager of the Advanced Electronics Department.

Abstract—Having discussed generally the SYNCOM, Early Bird, and the Applications Technology Satellite Systems in the two previous articles, we now turn to a more detailed investigation of power supplies for the satellite and balloon systems.

Solar-Cell Power Supplies for Weather-Balloon and Satellite Systems

THE successful operation of vehicles in space has been made possible, in part, by the availability of electric power. The sun has been the principal source of energy, and the silicon solar cell has provided the necessary conversion of photon to electrical energy. With few exceptions, such as the Echo balloon, or floating parts to be used for assembly of structures in space, all space vehicles without electric power become useless debris. The power subsystem is, therefore, one of the essential elements of every spacecraft, and its design often dictates the spacecraft configuration.

Power-level requirements, which continue to be below a kilowatt for a single mission, can be provided by solar cells within the packaging constraints of current boosters and payloads. Even multikilowatt power, however, seems practical with solar cells. Nevertheless, the increasing use of man in space and the desire for exploration of the distant planets provide a continuing motivation for the development of advanced power systems.

The two major areas of development are the electrochemical (H_2-O_2) primary fuel cells for the Gemini and Apollo spacecraft, and the large family of nuclear (SNAP) power systems. In the latter class, a thermoelectric system (SNAP-10A) has been launched and used. Several small radio-isotope-thermoelectric units (up to 5 W) have been successfully used in satellites, and the use of radioisotope units will grow in the future. Progress continues in solar-thermionics, but this type of power has yet to become identified as the power source for a specific mission or flight.

This article will discuss silicon solar-cell configurations for balloons and satellites. The following article by William Homeyer will consider isotope-powered thermoelectric generators as a possible satellite power source.

There are a number of critical constraints to be considered in the design of a weather-balloon power system. Environmental factors include a reduced atmosphere, a temperature range which may vary from —80° to +150°F, and high humidity. The most critical of these is temperature since it will affect operation as well as duration of flight. There is, however, another important criterion: weather balloons may operate in the vicinity of high-flying aircraft. The hazard to aircraft is not serious so long as the balloon is considered by itself. But any massive body, such as a block of electronic equipment, or a battery, will constitute a severe hazard to aircraft. It is imperative, therefore, that concentrated masses

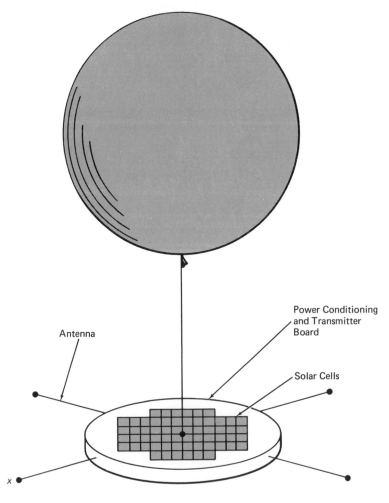

Figure 1   Weather balloon configuration.

be held to a minimum. In the power system, the three major units of concentrated mass are the solar array, the electronic components, and the batteries. In the following paragraphs, data will be presented for typical components applicable to the balloon mission. These components will be suitable for a 2-to-4-W power system in an intermittent 10-W broadcast power profile. This power system would be carried beneath the balloon as shown in Figure 1. The lightweight photovoltaic power source for the balloon mission can be made with thin solar cells bonded to an H film or a lightweight honeycomb substrate. The H-film substrate is superior from the standpoint of safety in case of aircraft collision, but it is more susceptible to accidental damage. The specific weight for the H-film array is approximately 0.25 lb/ft² utilizing 4-mil solar cells. The weight for a lightweight rigid substrate with 12-mil silicon solar cells is approximately 0.50 lb/ft². Figure 2 is an example of a lightweight H-film array designed and built for the GHOST program.

Figure 2 Sample of lightweight "H" film array.

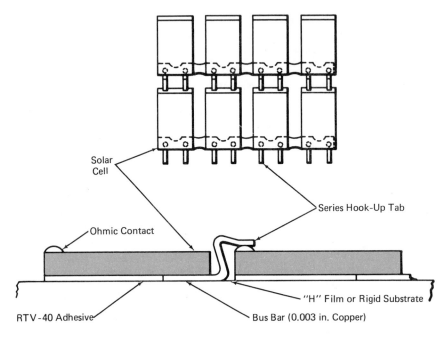

Figure 3   Submodule solar cell construction and connection.

Solar cells are generally assembled into submodules consisting of a number of cells in parallel, depending on the current requirements of the battery charger. Figure 3 is a schematic drawing of a module of four cells in parallel. These submodules are mounted on the substrate and connected in series to provide the necessary voltage for battery charging.

The components used in the fabrication of the solar array must be capable of surviving the rigors of atmospheric disturbances found at 30,000 ft, which include high winds, severe cold, rain, and ice. Solar cells have shown great resistance to degradation under severe environmental conditions such as encountered in high atmospheres. It is possible, but not necessary, to encapsulate the solar cells in transparent plastic to protect them from the environment.

Batteries will be required in both the balloons and satellites to supply peak power requirements. This energy-storage system may consist of conventional batteries, however, flexible tape batteries appear to have significant advantages for the balloon application.

Sufficient developmental work has been done to indicate that a flexible thin-film storage battery is feasible. Cells have been fabricated with a total thickness of 0.007 in. They consist of a positive electrode, a negative electrode, a 2-ply separator, and two encapsulating films, each of which has low permeability to liquids, gases, vapors, and light. Figure 4 is a diagram of such a battery.

These cells can be fabricated in any desired size and shape and connected in series to obtain the required voltage. The cell differs from conventional vented cells in that (1) it is completely sealed (which imposes limitations on charging rate), and (2) the small cross section of conducting elements limits the maximum current which can be delivered. In relatively low-current-density applications, such as the balloon application, neither of these limitations is expected to be significant.

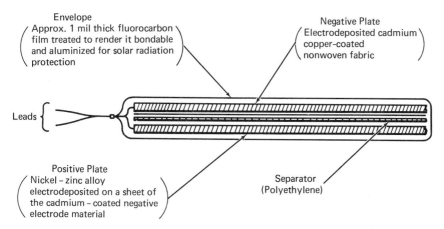

Figure 4   Diagram of a flexible thin-film storage battery for use on weather balloons. Note: Composite cell wetted with electrolyte solution (KOH).

The thin-film battery has an advantage over conventional nickel-cadmium batteries for low-temperature operation in that the electrolyte path is relatively short and of extremely large cross section. This means that internal resistance changes due to temperature are lower in the thin-film battery for a given voltage and current than would be expected from a comparable conventional battery. Limited work carried out with finishes having selective radiation absorption properties indicates that it will be possible to use the earth's infrared radiation to hold the actual internal temperatures of the system within much less severe limits than the natural temperature extremes of the atmosphere in which it is operating.

The method of deploying the battery in the balloon system has an important bearing on battery performance, particularly with respect to temperature extremes, gassing, moisture loss, and contamination by absorption of carbon dioxide. The battery could be deployed within the envelope of either the main balloon, or an auxiliary pressurized envelope. In either event, the fact that the battery is surrounded by an inert gas will materially reduce the chance that carbon dioxide or ammonia will "poison" the electrolyte. Operation within a balloon which will be inflated with a moisture-saturated, inert gas will tend to prevent bubble formation within the battery and will lessen the rate of moisture transfer out of the system. Finally, the presence of a gas at relatively high pressure outside the battery will permit convection current to equalize temperatures, and may even provide a means of maintaining a battery temperature significantly higher than the minimum temperatures reached by the surrounding atmosphere. This latter possibility would require the balloon walls to be a better absorber of infrared radiation than the atmosphere in which the balloon is floating, and would thus utilize the fairly intense and constant infrared radiation upward from the earth to warm the balloon and its contents. The warming effect produced in this manner will vary with the ambient temperature, ranging from a negligible value near 0°C to a number of degrees under the most severe low-temperature conditions.

The power conditioning electronics for the power subsystem must provide the necessary control and regulation of the power supplied by the solar cells and/or battery. The sophistication of the electronics is determined by the variation in input parameters (such as cell voltage and battery voltage), the voltage regulation required, and the efficiency required. The considerations below apply to the satellite as well as to the balloon batteries. To determine the variations in input parameters, it is necessary to examine the effect of solar-cell output as a function of temperature and of orientation, and the effect of battery voltage as a function of charge condition. The temperature affects the output characteristics of a solar cell as indicated in Figure 5. Note that the voltage of the cell is seriously affected, whereas only a small change in current occurs. Note also that the available output power of the cell changes with temperature. Power conditioning electronics must combine the solar array and battery characteristics appropriately and provide an output compatible with the equipment in order to use the electrical power. It must provide appropriate control of battery charging during sunlight hours and of battery discharge during the hours of darkness. If the using equipment does not require voltage regulation, the power subsystem can be quite simple, as illustrated in Figure 6(a). Diodes CR3 and CR4 are zener diodes that provide a voltage limiting function for the solar array so that battery charging will not be excessive. The zener diode voltage is set to equal the end-of-charge voltage of the battery. Diodes CR1 and CR2 prevent any discharge currents from flowing from the battery into the solar array

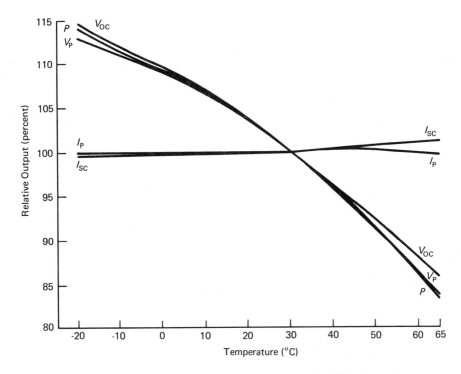

Figure 5   Plot of relative output versus temperature for N/P solar cells.

V_{oc} = open circuit voltage
V_p = maximum power voltage
P = power maximum
I_{sc} = short circuit current
I_p = maximum power current

during dark time operation. The zener diode limiters (shunt regulators) must be capable of dissipating the "excess" power provided by the solar array. The amount of power can be determined by examining the solar-array output power at the zener voltage and the particular conditions the array will experience (such as temperature and orientation with respect to the sun) and subtracting from this value the power supplied to the battery.

If output voltage regulation is required a series regulator can be added to the output as shown in Figure 6(b). The sophistication of the series regulator will be determined by the output voltage tolerance required. The efficiency of the regulated system is lower than the unregulated system owing to the additional power losses associated with the series regulator.

It is possible under certain conditions that the battery charge rate will

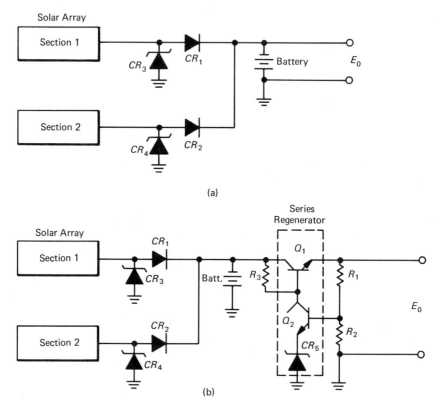

Figure 6   Power conditioning electronics. (a) Without output regulation; (b) with output voltage regulation.

be so excessive that the battery will be damaged. This can happen if the solar array is cold and the orientation with respect to the sun is too "favorable." Careful attention must be paid to the battery characteristics and its ability to accept high charge rates. If it is determined that this condition cannot be avoided, a charge current limiter must be added to the system.

The satellite power system for a meteorological balloon program will be determined by the data handling requirements. The amount of data which must be collected and retransmitted is a function of the information to be obtained from each balloon as well as the number of balloons used.

Based upon previous studies a 200-W satellite power system will be in the range of interest for a meteorological balloon application. This power system is capable of intermittent operation at higher power levels when it has been commanded to relay considerable accumulated data. During interrogation of balloons it is expected that nominal power, in the range of 50 to 150 W, will be required.

In the consideration of various solar array configurations, several basic patterns and designs may be chosen. Some of these are shown pictorially in Figures 7, 8, 9, 10, and 11.

Figure 7   Sun oriented solar array.

With regard to the flexible array (Figure 10), the state of the art may not be sufficiently advanced to be considered reliable. The main advantage of this design is high packaging density, particularly suited for carrying large arrays in confined spacecraft.

The curved, rigid panel is an interesting concept, combining rigidity and high packaging density with low specific weights. This type of panel may be a descendant of the flat panels described below.

The rigid, sunlight-concentrating panel design (for example, the V-ridge panel) is structurally quite good and is similar to the Jet Propulsion Laboratory Mariner IV design. However, it does suffer certain disadvantages, as follows:

(1) slightly larger area, compared with a flat panel (normal to the sun) for a given power load,

(2) possible large loss in power due to reflective surface degradation,

(3) higher cell temperature (76° vs. 55° C for a flat panel), and

(4) greater power loss vs. misorientation.

The sun-oriented, flat panel, shown in Figure 7, has size and weight advantages; however, some critical areas exist in its integration with the vehicle. Since the vehicle attitude control may be used to aim the antenna system, a separate positioning system would be required for the solar array. Thus, bearings would be needed to allow rotation of the array, and slip rings would be required for power transfer.

The semioriented paddles, shown in Figure 8, have been used on many

Figure 8   Semisun oriented solar array.

Figure 9   Nonoriented, body-mounted solar array.

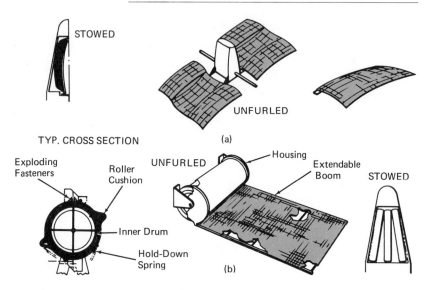

Figure 10   (a) Biconvex panels; (b) flexible rollout solar array panels.

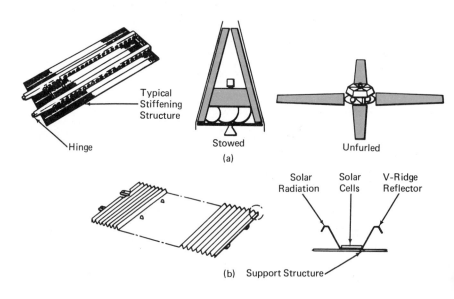

Figure 11   (a) Flat panels; (b) concentrating solar array panels.

spacecraft such as IMP, OAO, and Explorer XII, and have proved feasible. A major problem with such paddels is shadowing of portions of the solar array by other paddles or by the spacecraft itself, which results in a power loss. Two electrical design methods are available to minimize these losses: laying out the cells to insure that only small sections are lost to shadowing, or using switching mechanisms to take shadowed sections out of the electrical circuit.

The nonoriented, body-mounted array shown in Figure 9 has been used successfully in the SYNCOM series and Early Bird satellites. The major problem with this array is low utilization of the total solar cell active area and possible temperature variations across the array. However, for low-power ouputs, this array is the simplest to incorporate into a space vehicle. Assuming 10 W/ft² for an oriented array, a nonoriented, body-mounted structure requires approximately fives times the total area of the oriented array, or equivalently, the specific power is 2 W/ft². For a solar array power of 200 W, 100 ft² of surface area is required. Table 1 summarizes the area required to supply 200 W using a sun-oriented, a nonoriented, and a nonoriented, body-mounted array.

TABLE 1
REQUIRED AREA FOR A 200-W SOLAR ARRAY

	Sun-Oriented Panel	Nonoriented Panel	Nonoriented, Body-Mounted Array
Area	20 ft²	65 ft²	100 ft²

For the first six years of the space age, most space vehicles used P/N silicon solar cells. The selection of these cells was based on availability and performance characteristics determined at that time. Now, however, virtually all the solar-cell manufacturers have converted their production lines over to N/P cells rather than P/N cells, and the major developmental emphasis is placed on improving the N/P silicon cells. Solar cells of the N/P type with dimensions of 2×2cm with a nominal thickness of 0.013 in. and a nominal weight of 0.03 g have found the most widespread application.

The space environment of the satellite consists of a near vaccum (about 10^{12} torr at 7000 nmi) containing charged and uncharged particles; electromagnetic radiation from the sun, earth, and other bodies; meteoroids; and electric, magnetic, and gravitational fields. The following paragraphs briefly summarize the magnitude and effects of the charged particle and meteoroid environment; these two areas constitute the greatest hazard to power systems operation.

In solar flares, the abundance of constituent particles, energy spectra and range, and the total particle flux, vary widely with each flare event to an extent depending on the size of the flare. The most serious component of the particle radiation is the high-energy protons.

The tempo of fluctuation in the energy spectra at each event is rapid, sometimes reaching a peak in a few hours, and then decaying in a manner approximately proportional to T^N in a few days. Solar-flare events ($N \approx -3$) follow an approximate 11-year cycle, minimum activity being in the years 1964–1965 and the minimum activity in 1969–1971. In 1956, a year of near-peak activity, the integrated intensities of solar particles (protons) above 30 and 100 meV received at the earth's distance from the sun were 8×10^9 and 8×10^8 particles/cm², respectively, with two major storms. The time-integrated proton flux per year for $E > 30$ meV and $E > 100$ meV is 1×0^{10} and 9.6×10^8 protons/cm²/yr, respectively. The spectral distribution of the solar-flare protons is obtained from a power-law representation utilizing the above data for 30- and 100-meV protons.

The geomagnetic field shields a satellite at altitudes below approximately 15,000 nmi and 45° inclination from solar-flare effects, although some small degradation can be anticipated. At synchronous altitudes, Van Allen radiation decreases and the solar flare degradation assumes major importance.

The Van Allen belt radiation consists of energetic particles (protons and electrons) trapped in the earth's magnetic field. Analysis of data obtained from a series of exploratory satellites reveals the following:

(1) The radial distance to the outer boundary of the geomagnetic trapped radiation varies, ranging from 8.5 to 11.0 earth radii.

(2) Electrons are more or less uniformly trapped between two earth radii and the outer boundary.

(3) Protons are trapped throughout the same region.

Unfortunately, specific data are not available at present on such variables as the differential energy spectrum and temporal fluctuations at all altitudes and inclinations of orbits under consideration. Figures 12 and 13 show the Van Allen belt radiation for electrons and protons, respectively.

The degradation due to particulate radiation (solar flare protons and/or Van Allen belt radiation) is a function of the protection given the solar cells by glass coverslides bonded to the surface of the solar cells and the

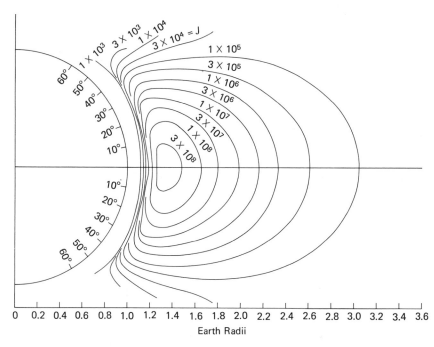

Figure 12   Electron density map for the Van Allen Belt (electrons/cm²-sec).
Note: Energy > 0.5 MeV.

critical flux ϕ_c. The critical flux is defined as the integrated particle flux necessary to cause a 25 percent decrease in the initial output power of a solar cell. The critical flux of a solar cell is a function of the energy of the impinging particle. However, in order to simplify calculations involving the radiation degradation to solar cells, one critical flux for a particular energy range is used to determine power degradation. The critical flux ϕ_{cp} for protons is typically 2×10^{11} protons/cm² for 12-mil silicon solar cells and is approximately 5×10^{11} protons/cm² for 8-mil silicon solar cells over the energy range of 1 to 5 meV.

The presence of a glass cover-slide filter (cover glass) on a solar cell serves only to cut off lower-energy particles and thus decrease the incident flux that reaches the surface of a solar cell. An example of the effect of a cover glass is as follows.

A 25-mil quartz cover glass has a proton energy cutoff of 10 meV. Particles with energies greater than 10 meV will penetrate to the cell surface and into the bulk of the solar cell. For a 25-mil cover slide, this means that protons in the energy range of 11 to 16 meV will cause most of the radiation degradation, since the proton flux above 16 meV is small, relative to that below 16 meV.

The power degradation caused by particulate radiation, especially in the trapped radiation belts, that is, Van Allen belts, is due to both electrons and protons, whereas in interplanetary space, or at synchronous altitudes, radiation damage is primarily due to solar-flare protons. However, the radiation degradation due to protons (and that due to electrons) is not linearly additive. This damage is taken into account by determining the degradation due to electrons when made equivalent to protons. The equivalent number of electrons, that cause the same percentage degradation as one proton, has been determined experimentally to be from 3000 to 10,000. Therefore, it is necessary to divide the incident electron flux by

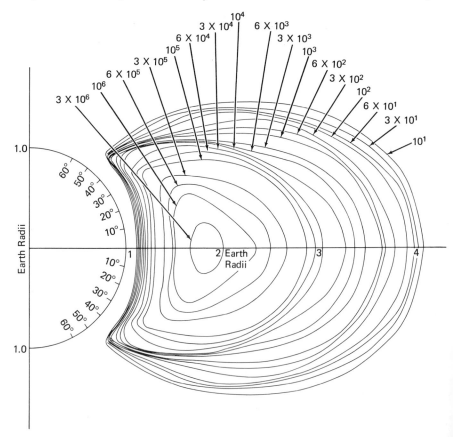

Figure 13   Proton density map for the Van Allen Belt (protons/cm²-sec). Note: Energy > 4.0 MeV.

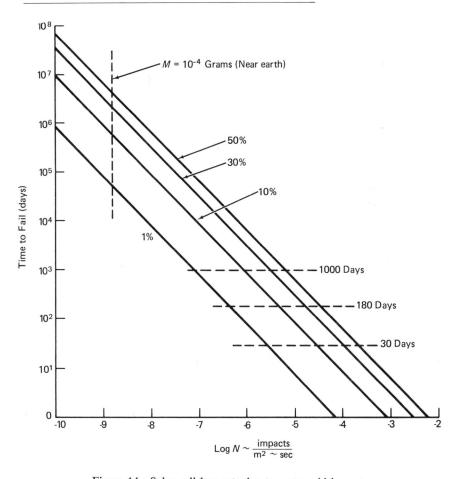

Figure 14   Solar cell loss rate due to meteoroid impacts.

a conservative factor of 3000 to obtain the equivalent proton flux. This flux is then added to the incident proton flux to obtain the total incident equivalent proton flux used in the determination of the power degradation. The factor of 3000 is a typical average value for the N/P cells available today.

The typical effect of meteoroid bombardment on solar cells is shown in Figure 14, which gives solar cell losses as a function of bombardment time, assuming one cell is lost per impact.

This figure is derived using a near-earth micrometeoroid model of

$$\log N_E = -13.8 - \log M$$

where N_E is number of particles/m²·sec of mass M and greater, moving at a distance of 1 AU from the sun (that is, near earth space).

The assumption that one cell is lost per impact is a worst case assumption, since a micrometeoroid impacting a surface will generally damage only a small area of the impacted surface. In an impact on a solar cell, a small segment of the active area will be lost, but the cell is still capable of producing usable power. The impact of a large meteoroid can cause catastrophic failure in the entire array, but the probability of this occurrence is so infinitesimal that it is assumed to be zero. It would require over ten years in an earth orbit to lose but 1 percent of the total cells on the array, assuming one cell lost per impact. It can be assumed, therefore, that power degradation due to meteoroid bombardment is negligible during the time period of most missions.

The batteries used in the satellites would be chosen from nickel or silver cadmium or silver zinc, subject to the following general requirements. The total wet life of the battery should be at least 450 days in orbit plus two or three months from date of manufacture to launch. Temperatures may range from —65° to 125°F. Recharge times will depend on the number of times per day the satellite transmits its data to earth tracking stations. This in turn depends on the number of balloons being monitored by the satellite and the nature and urgency of the data being collected.

Full investigations of all the parameters involved in the choice of satellite batteries are available, but would be too detailed for this book. However, some of the parameters considered in more detail for the balloon hold for the satellite batteries as well.

WILLIAM HOMEYER

William Homeyer studied chemical engineering at Rensselaer Polytechnic Institute and received his M.S. and Ph.D. degrees in nuclear engineering from the Massachusetts Institute of Technology. He has been working for the past six years for the General Atomic Division of General Dynamics Corporation in San Diego on the development of thermoelectric and thermionic power supplies for space applications.

Abstract—To date, as Mr. Winsor suggested in the previous article, the silicon solar cell has been the principle source of power for space vehicles. However, as power level requirements continue to increase and as satellites become larger and more sophisticated, electric power supplied by nuclear sources becomes increasingly attractive. Dr. Homeyer concludes Part 4 by discussing heat sources, the energy converters, and the heat rejection systems of several isotope nuclear power systems.

Nuclear Power Supplies

NUCLEAR power supplies for space applications have been under development since the early days of space satellites; however, there have been very few applications in which they have been used so far. The high energy density characteristics of nuclear systems makes them ideally suited for space power, but there are many development problems that have been very difficult to solve because of the hostile environment of space and the requirements for high temperatures, so that waste heat can be rejected efficiently by radiation. All the nuclear-powered devices that have been used in space so far have employed direct conversion of heat to electricity without an intermediate mechanical energy stage. This paper will look briefly at the entire field of nuclear power in space, and then concentrate on isotope-powered thermoelectric generators, which are most directly applicable to weather-observation satellites.

The components of any nuclear space power system are a heat source, an energy converter to convert the heat to electricity, a heat rejection system, and in some cases, a shield to protect components or people (for manned applications) from radiation, and an additional optional component of power conditioning equipment which is frequently required to match the electrical output of the generator to the load.

Two types of nuclear heat sources have been under development. These are reactors and isotopes. The reactors offer very high power densities and can supply a large quantity of power at a relatively low weight. They have size limitations set by nuclear criticality, however, so that they are probably not applicable economically below something on the order of 10 kW of electrical output. Isotopes have essentially lower thermal power densities and higher weights for a given power, but there are no size limits, and in fact, isotopic heat sources that produce just over a watt have been developed for special terrestrial applications. Two basic types of isotypes have been used. These are alpha emitters and beta-gamma emitters. Alpha emitters have the very important advantage that there is very little ionizing radiation and only for very large blocks of power

121

is any shielding required. Beta–gamma emitters, on the other hand, have a very high level of ionizing radiation from the penetrating gamma rays and from *bremsstrahlung* from slowing down of the beta particles. The alpha emitters are relatively cheap because they are waste products from nuclear reactors. The isotopes which have been considered as heat sources are shown in Table 1.

TABLE 1
ISOTOPES FOR HEAT SOURCES

α emitter	Pu^{238}	Po^{210}	Cm^{242}	Cm^{244}
Half-life	86 years	138 days	163 days	18 years
Specific power (W/g)	0.5	140	11	3
β–γ emitter	Sr^{90}	Ce^{144}	Cs^{137}	
Half-life	28 years	285 days	30 years	
Specific power (W/g)	0.7	20	0.4	

The alpha emitters are listed on top and the beta-gamma emitters beneath. The half-life, or the time that it takes for half of the isotope to decay is listed below each isotope. Notice that the half-lives of the isotopes of interest are in the range from 100 days to 100 years, which are the boundaries of the mission times for which isotope-powered generators are being considered. Any isotope with a half-life of 100 days or less could be used only on very short missions on the order of a month or two. For example, polonium 210, which is a very interesting isotope, is limited to about 3 months of practical mission lifetime because in 138 days one-half of the initial charge of fuel would be gone. Plutonium 238, on the other hand, can be used for very long missions, since fuel consumption is less than 1 percent per year. Notice that the half-lives and the specific powers are inversely related to one another, that is, we can have very high specific powers and low fuel weights with a short lifetime or much lower specific powers and higher weights with a long-lived isotope. The three isotopes that have been considered most seriously are plutonium 238, which has been used in satellites that have flown, polonium 210, which has been used in demonstration generators and is being considered for the next generation of satellites, and strontium 90, which is being considered seriously but has not yet been used because of its intense radiation and resultant safety hazards.

Several different types of energy converters have been considered so far. The turboelectric and thermionic converters are efficient in the high power range, and are thus associated with the reactor power systems. For the lower power systems, various isotopic thermoelectric converters have been designed.

The two turboelectric systems under development for use in space are based on the Rankine cycle and Brayton cycle. The Rankine cycle is the cycle of the steam plant in which the working fluid vaporizes, expands through a turbine, and is condensed; and the Brayton cycle is the gas turbine cycle in which a gas is compressed, heated, expanded through a turbine, and then cooled and recompressed.

In space, the major problem of converting heat to electricity is that waste heat must be rejected from the power plant at a relatively high temperature in order for the radiator to be of a reasonable size. This is because heat can be transmitted in space only through thermal radiation and the radiant heat flux varies as the fourth power of temperature. As a result of the high heat rejection temperature, efficiencies are generally low, both turboelectric and thermionic systems being on the order of 10 to 20 percent. The working fluids that have to be considered for the Rankine cycle are metal liquid-vapor systems such as mercury and potassium. Both these and the gas systems have problems of leakage into space; there can be very little leakage of the working fluid from the conversion system if it is to operate for long periods of time without periodic addition of working fluid.

A simplified thermionic conversion system is shown in Figure 1. This is an in-core or reactor thermionic system in which a fuel material is heated by nuclear fission, which in turn heats a tungsten emitter. Electrons are emitted from the tungsten emitter and travel across the interelectrode space to the collector, whose temperature might be as high as 1000°C, as shown here, or as low as 700°C, which is about the optimum for efficiency. The heat would then be removed from the outside of the collector by a circulating liquid metal coolant. Cesium vapor at a pressure in equilibrium with liquid cesium at about 300°C is present in the interelectrode space to neutralize space charge and modify the electrode work functions. A reactor made up of thermionic cells has the advantages over the turboelectric systems that it involves no moving parts in the conversion system and requires no seals of rotating shafts in space. It also is characterized by a very high heat-rejection temperature at which any of the turboelectric systems would be inoperative.

Figure 2 is a diagram of a thermoelectric conversion system. Heat flows from source to sink through the N and P elements, producing a voltage gradient and hence a flow of current through the elements. The two materials that have been considered for space applications and have been devel-

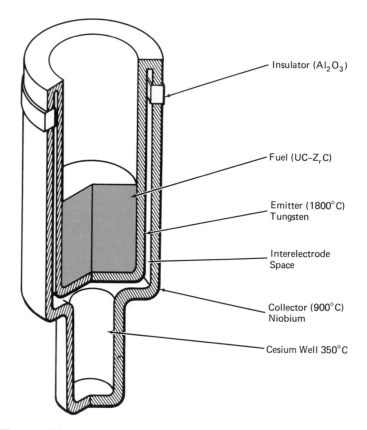

Figure 1   Nuclear thermionic cell (schematic drawing).

oped are lead telluride (or lead tin telluride) and germanium–silicon alloy. The germanium–silicon alloy tends to have a lower efficiency, on the order of 2 or 3 percent, as compared with 5, 6, or 7 percent for the theoretical limits on the lead telluride, lead tin telluride system. The germanium–silicon alloy, on the other hand, can operate and reject heat at a higher temperature so the radiator requirements are lessened.

Two basic generator designs are shown in Figure 3: the pressure contact and the metallurgically bonded designs. The pressure contact design has been used in all the isotope-powered thermoelectric generators that have been flown to date. In this design large thermoelectric elements are pressed against metallic contacts by springs which maintain the elements in compression. The elements are surrounded by thermal insulation to reduce the heat bypass loss. These devices tend to have very high heat fluxes and require fins or other extensions of the radiator area for cooling.

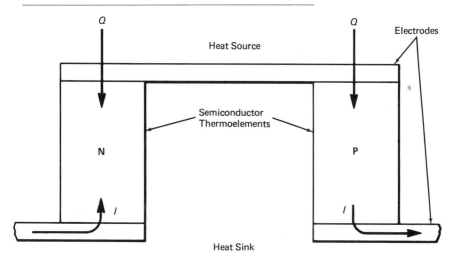

Figure 2 Schematic of mode of operation of P/N semiconductor thermo-electric system.

The other approach shown in the bottom of the figure is to use small elements which are metallurgically bonded to metallic hot and cold shoes and are strong enough to serve as structural members. The collector structure shown is also the hot-side electrical connector and is supported by the elements attached to the cold-side honeycomb support structure and to aluminum cold-side connectors to complete the electrical circuit. These panels are sized to reject heat directly from the cold side by thermal radiation so that no additional radiator surfaces are required. They receive heat on the hot side by thermal radiation which reduces the problem of differential thermal expansion. Details of the thin panel design are shown in Figure 4. A honeycomb sheet is the basic structural member on the cold side, a thin multilayer reflective material which is a good thermal insulation in vacuum is above this, and finally the collector on top is held in place by the thermoelectric elements.

The operational generators that have flown so far are summarized in Table 2. These are the SNAP-3 and SNAP-9A, which employ plutonium 238, and the SNAP-10A, which is a reactor system using the uranium zirconium hydride compact reactor. Note that the power production varies over several orders of magnitude for $2\frac{1}{2}$ W for the SNAP-3 generator, the first isotope thermoelectric generator that was flown, all the way to 500 W for the reactor-powered system. The thermoelectric material used in the isotope systems is lead telluride, which has a higher efficiency. Germanium–silicon alloy is used in the reactor where efficiency is not important, since the reactor could produce many times as much heat with-

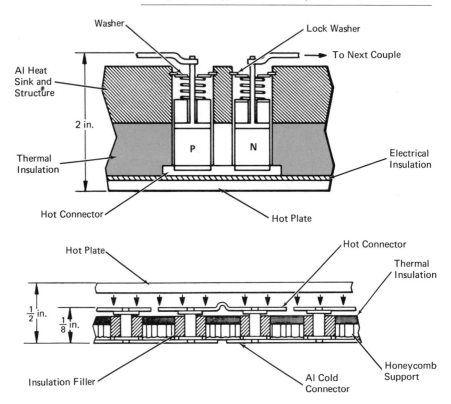

Figure 3 Pressure contact and metallurgically bonded thermocouple designs.

out a change in size. Note that the reactor performance, in terms of watts per pound, is little more than half that of the isotope system. This is because the output power of 500 W is too low for the reactor, which has a minimum size, to be used economically. The weight of the isotope system (SNAP-3 and SNAP-9A) varies nearly linearly with the power for given choices of isotope and converter, so that the watts per pound is nearly constant.

Development work is in progress to perfect the technology of thin, light-weight converter panels with bonded lead telluride elements. This approach can reduce the weight of the converter by an order of magnitude. For a heavy isotope like plutonium 238, this can bring the performance of the generator up to 2 W/lb as compared with 0.8 W/lb for the present generators (Table 2). For a light-weight isotope like polonium 210, for shorter time missions, still greater improvements in performance are possible.

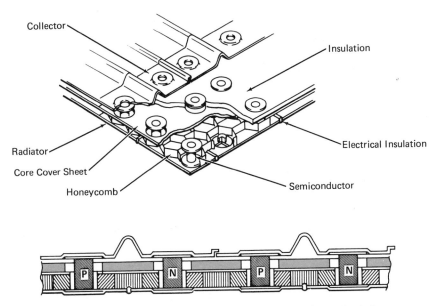

Figure 4   Details of the thin panel (metallurgically bonded) design.

There is also work under way to raise the efficiency of thermoelectric conversion, which now stands at about 5 percent for the lead telluride system. The technique to be used in raising efficiency is to cascade two converters using germanium silicon alloy as the high-temperature stage, which accepts heat from the heat source at 800° to 1000°C and rejects the heat at 400° to 500°C to a second panel of lead telluride–lead tin

TABLE 2
OPERATIONAL NUCLEAR POWER UNITS

	SNAP-3	SNAP-9A	SNAP-10A
Heat source	Pu-238 Isotope	Pu-238 Isotope	U-ZrH Reactor
Power conversion	Thermoelectric PbTe	Thermoelectric PbTe	Thermoelectric GeSi
Electric power	2.5 W	25 W	500 W
Efficiency	5 percent	5 percent	1.8 percent
Weight (unshielded)	3 lb	28 lb	950 lb
Area	0.28 ft²	2.8 ft²	63 ft²
Specific weight (unshielded)	0.8 W/lb	0.9 W/lb	0.5 W/lb
Specific area	9 W/ft²	9 W/ft²	8 W/ft²

telluride operating in the temperature range from 350° to 450°C down to 150° or 200°C.

The relation between temperature difference across the converter and isotope weight is worth noting. To keep the specific weights (in watts per pound) and hence the efficiencies of the heavier isotopes competitive requires a large temperature difference across the conversion elements. For the lighter isotopes, a sacrifice in efficiency allows a higher radiator temperature and hence a lighter conversion system. For any set of components there is an optimum balance between the weights of the converter and the fuel.

TABLE 3
*WEIGHT ANALYSIS OF THE VARIOUS TYPES OF POWER
SOURCES FOR VARIOUS UNMANNED MISSIONS*

		0.5 ekW	5 ekW	50 ekW	500 ekW	5000 ekW
Reactor thermionic	Reactor and conv.		1000 lb	1000 lb	3200 lb	13,000 lb
	radiator		10	100	1000	10,000
	shield		400	650	1520	4,000
			1410	1750	5720	27,000
SNAP reactor thermoelectric	Reactor radiator and conv.	350 lb 92	350 lb 920	950 lb 9200		
	shield	300	600	1500		
		742	1870	11,650		
Alpha isotope thermoelectric	Isotope structure, conv.	26 lb 60	264 lb 600	2640 lb 6000		
	shield	4	12	72		
		90	876	8712		
Gamma isotope thermoelectric	Isotope structure, conv. shield	118 lb 60 3	1180 lb 600 10	11,800 lb 6,000 50		
		181	1790	17,850		
Solar cells earth orbit	High orbit					
	Oriented	73 lb	730 lb	7,300 lb		
	Unoriented	270	2700	27,000		
	Low orbit					
	Oriented	172	1720	17,200		
	Unoriented	640	6400	64,000		

The major problems in application of isotope power supplies have been and are in safety. All these isotopes are deadly if swallowed or inhaled and therefore must be encapsulated in some manner so that they cannot possibly get out. "Cannot possibly" becomes a major problem when considering a rocket launch with a destructive abort possible at any stage in the flight. The technique with the alpha emitters has been to seal them in a capsule with a large internal void volume for accumulation of helium. The helium, which is a product of the alpha decay, can produce a considerable pressure unless an adequate void volume is provided. The beta–gamma isotopes require shielded shipping containers to protect against their ionizing radiation. These isotopes also require great care in remote handling to place them in the generator in a rocket on the launch pad.

Table 3 gives an overall comparison of the weights of the various systems at several power levels. The weight estimates, except for the SNAP-10A reactor, are based on optimistic estimates of what might be built two or three years from now, not on present technology. Notice that the solar cells vary by almost a factor of 10 from 73 lb for oriented panels in a high orbit, where they are in the sun most of the time, to 640 lb for unoriented panels in a low orbit, where they are in the shade more of the time and are frequently turned away from the sun.

Table 4 gives a comparison of the areas for the different systems. Notice that the isotope thermoelectrics are quite compact by comparison with the solar cells, while the reactor thermoelectrics are on the same general order, but a bit more compact for a given efficiency. The reactor ther-

TABLE 4

AREA ANALYSIS OF THE VARIOUS POWER SUPPLY SYSTEMS (ft^2/ekW)

Solar	high earth orbit		oriented	110 ft²
photovoltaic			unoriented	400 ft²
	low earth orbit		oriented	190 ft²
			unoriented	700 ft²
Isotope	radiator temperature			
thermoelectric	10	percent efficiency	150°C	67 ft²
	9	percent efficiency	200°C	48 ft²
	7.5	percent efficiency	250°C	38 ft²
SNAP reactor	radiator temperature			
thermoelectric	5	percent efficiency	325°C	40 ft²
	3.4	percent efficiency	325°C	70 ft²
Reactor	radiator temperature			
thermionic	10 percent efficiency		900°C	1 ft²

mionic systems, on the other hand, are extremely compact because of
their very high heat-rejection temperature.

TABLE 5

COST ANALYSIS OF POWER SUPPLY SYSTEMS ($/ekW)

Solar cells	high orbit	oriented	$ 440,000
		unoriented	1,600,000
	low orbit	oriented	$ 760,000
		unoriented	2,800,000
Alpha isotope	Pu-238		$6–12,000,000
	Cm-244		6–12,000,000
	Po-210		620,000
Gamma isotope	Sr-90		$ 250,000

Table 5 is a comparison of the costs in dollars per electrical kilowatt
among solar cells and various nuclear power systems. Notice the large
difference in isotope cost between the plutonium 238 and curium 244
and polonium 210. The reason for this is that the plutonium 238 and
curium 244 are transuranium elements produced by irradiation of uranium
and have to be separated from uranium and other transuranium elements.
The separation has to be frequent, and a radiation schedule has to be
followed carefully in order to keep the plutonium and curium from being
too badly contaminated with plutonium 239. Polonium 210, on the other
hand, is obtained comparatively easily by irradiating bismuth in a reactor.
The polonium product is removed from the bismuth target by a simple
chemical separation. Strontium 90 is relatively inexpensive because it is
a by-product in processing spent fuel from nuclear reactors.

operative weather satellite systems

ABRAHAM SCHNAPF

Abraham Schnapf studied mechanical engineering at the City College of New York and received his B.S. degree in 1948. After two years with the Goodyear Aircraft Corporation, he came to work for his present employer, Radio Corporation of America. In 1953, while working in the RCA Airborne Systems Department, he studied for and received his M.S. degree in mechanical engineering from Drexel Institute of Technology. Since 1958, Mr. Schnapf has worked on the TIROS weather satellite system and is the Project Manager of the TIROS/TOS Project at the RCA Astro-Electronics Division in Princeton, New Jersey.

Abstract—*The material in Part 5 follows closely that of Part 4. Part 4 contained an introduction to the various components of a general satellite system, whereas Part 5 deals directly with operational weather satellites. Mr. Schnapf begins with a discussion of the design and of the prolific returns of the TIROS television satellite system. TIROS has been the grandfather of weather satellites and has provided the first daily photographic coverage of the entire globe. This program has provided 10 research and semioperational meteorological satellites since 1960. In February 1966, the TOS (TIROS Operational System) program placed the ESSA satellite into orbit. To date, five ESSA satellites have successfully operated in orbit, providing continuous routine daily global weather observations to the United States and to meteorologists throughout the world. The second-generation TIROS Operational Satellite is now under development.*

The TIROS Weather Satellites

T IROS, the world's first weather satellite system, was initially planned as a research and development program to determine the feasibility of observing the earth's cloud cover from an earth-orbiting satellite. Since the time of the first launch on 1 April 1960, TIROS has performed beyond all expectations, and its role, consequently, has been expanded to perform routine global weather observation on a daily and continuous basis. In the first quarter of 1966 when the TIROS Operational System (TOS), built by RCA for the Environmental Science Services Administration (ESSA), began operations, it provided the first daily weather pictures of the entire world for the United States, while also providing local direct real-time television pictures to simple APT receiving stations located in many countries throughout the world.

In this article some of the technical accomplishments of the TIROS I through X satellites and the TOS (designated ESSA 1, 2, 3, 4, and 5) satellites will be reviewed, along with several aspects of the second-generation satellite programs and some ideas for the future of the weather satellite program.

Figure 1 presents a series of three pictures indicating the various missions that TIROS could fulfill. The first picture (upper left) depicts the orbits of the first eight TIROS satellites, which were launched between a 48° and 58° inclination owing to Delta booster launch limitations from the Eastern test range. The upper right picture shows the "wheel" mode that was introduced with TIROS IX. This satellite, launched in January 1965, was the forerunner of the TOS system and was placed in a near-polar orbit by means of a Delta 3C launch vehicle. With the use of the near-polar orbit and by maneuvering the satellite's spin axis from its orbit injection attitude to one in which the spin axis was normal to the orbital plane, it was possible to achieve global observation on a daily basis. The lower picture represents a simulated synchronous mission by placing a

133

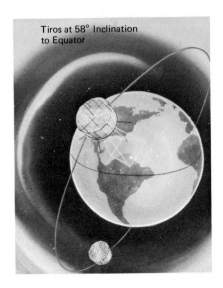

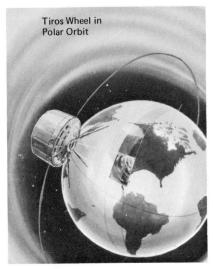

Figure 1   Present and proposed orbits for TIROS satellites.

spacecraft in a highly elliptical orbit with an apogee of 22,000 mi and a perigee of 300 mi. With a single TV photograph one could then observe the entire earth's disc.

Table 1 summarizes the various configurations and the important orbital parameters of the TIROS and TOS satellites. The weight of the satellite has ranged from 260 to 300 lb. Initially, the satellite system contained a narrow-angle camera with a 12° lens and a wide-angle camera with

TABLE 1
TIROS/TOS CONFIGURATION SUMMARY

	TIROS											ESSA			
Spacecraft designation	I	II	III	IV	V	VI	VII	VIII	IX	X	1 (TIROS)	2 (TOS)	3 (TOS)	4 (TOS)	5 (TOS)
Spacecraft weight	263	278	285	286	287	281	299	260	300	290	300	285	326	285	327
Orbit data															
Apogee (miles)	466	453	506	525	604	442	404	468	1602	521	523	879	923	894	883
Perigee (miles)	429	387	461	441	366	425	386	435	435	467	433	840	860	823	840
Inclination (degrees)	48.4	48.5	47.5	48.3	58.1	58.3	58.2	58.5	96.4	98.6	98	101	101	102	102
TV CAMERAS															
1	NA	NA	WA	MA	MA	MA	WA	WA	WA	WA	WA	APT	AVCS	APT	AVCS
2	WA	WA	WA	WA	WA	WA	WA	APT	WA	WA	WA	APT	AVCS	APT	AVCS
Sensors															
Wide angle IR		X	X	X											
5-channel IR		X	X	X											
Omni-directional IR			X	X			X								
Electron temp. probe							X								
Flat plate IR													X		X
Mission mode															
Axial	X	X	X	X	X	X	X	X							
Wheel (sun synchronous)									X	X	X	X	X	X	X
Attitude control															
MBC		X	X	X	X	X	X	X	X	X	X	X	X	X	X
QOMAC									X	X	X	X	X	X	X
MASC									X	X	X	X	X	X	X
Mech. damper	X	X	X	X	X	X	X	X							
Liquid damper															

NA = Narrow angle (12.5° lens) − 5000 mi²
MA = Medium angle (78° lens) − 250,000 mi²
WA = Wide angle (104° lens) − 500,000 mi² } 400 mi Altitude
APT = Wide angle (108° lens) − 1,000,000 mi²
APT = Wide angle (108° lens) − 4,000,000 mi² } 860 mi Altitude
AVCS = Wide angle (108° lens) − 4,000,000 mi²

MBC = Magnetic bias control
QOMAC = Quarter orbit magnetic attitude control
MASC = Magnetic spin control

(a)

Figure 2 The TIROS satellite. (a) standard configuration; (b) wheel configuration.

a 104° lens. This was later changed to two wide-angle lenses on TIROS III and a wide- and a medium-angle lens on TIROS VII. On TIROS VIII, the first APT (Automatic Picture Transmission) camera system was incorporated. This will be discussed later in more detail.

There were two schools of thought in the U. S. Weather Bureau concern-

(b)

Figure 2   (*Continued*)

ing the required picture resolution. There were those who sought high-resolution data and those who wanted to get the global picture; hence, the camera optics were varied from narrow- and wide-range to medium-angle and back to wide-angle lenses during the research and development phases of TIROS. Currently, the most useful operational information comes from the wide-angle systems, because of global coverage capability. The change in transmitting frequencies, lens configuration, and various infrared experiments, as well as the orbit inclinations, have been the main alterations to the TIROS I through VIII series of satellites. As mentioned earlier, the early orbits were inclined at 48° and later 58°, whereas the last seven satellites utilized near-polar, sun-synchronous orbits. TIROS X employed a sun-synchronous orbit, with an orbital drift of about 2° per year.

Figure 2 shows the two key configurations in the TIROS program. On the axial system, as in the TIROS I through VIII and TIROS X configurations, the cameras were oriented parallel to the spin axis. The satellite was launched with the spin axis in the plane of the orbit; hence, the cameras were able to observe only the sunlit earth approximately 25 percent of the time. With the wheel-oriented, polar-orbiting satellites (TIROS IX and ESSA 1, 2, and 3), the cameras were mounted radially or normal to the spin axis, and the satellite was maneuvered into an orbital position so that the spin axis was oriented normal to the orbital plane. Thus, with each rotation of the satellite about its spin axis, the camera was able to observe along the local vertical; and with the sequential triggering of the camera shutter at the prescribed intervals, observation of the entire sunlit earth was possible during each day. Figure 3 shows the basic orbit for both the standard TIROS and the TIROS wheel satellites. By means of the reorientation of the cameras with the wheel satellites and by using refined magnetic stabilization techniques, it was possible to provide global coverage daily and to increase the photo-coverage by a factor of four over the earlier axial system.

Figure 4 shows the key subsystems that make up a standard TIROS

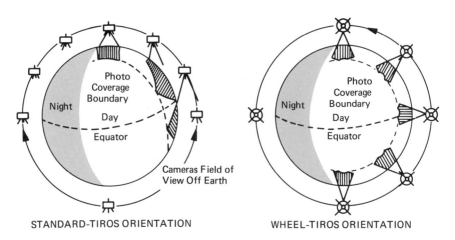

STANDARD-TIROS ORIENTATION   WHEEL-TIROS ORIENTATION

Figure 3   Orientation in orbit of TIROS satellites.

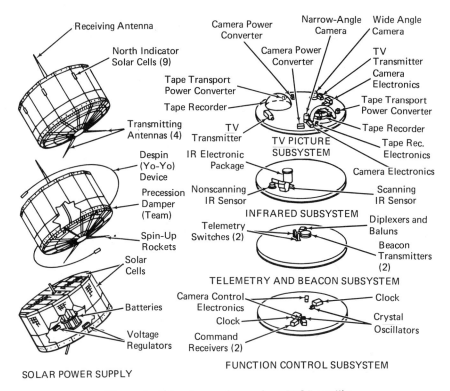

Figure 4   Subsystem makeup of a TIROS satellite.

satellite. There are roughly 10,000 solar cells around the sides and top of the satellite to provide power. The lower left picture shows more detail of the power supply. The solar cells convert solar energy into electricity to recharge the 63 nickel-cadmium batteries during the daylight part of each orbit. The whip antenna receives 148-MHz command signals from a TIROS Command and Data Acquisition (CDA) station on a 30-MW receiver. The crossed-dipole transmitting antenna on the underside of the satellite transmits the television data at 235 MHz and the telemetry data at 136 MHz.

The satellite is put into orbit spinning at 125 rpm, owing to the requirements of the third-stage solid rocket. The satellite is despun after separation from the third stage by the deployment of a "yo-yo," or a weighted system. This system is unleashed by a despin timer and unwraps itself from the periphery of the satellite because of the centrifugal force acting on the weights and detaches from the satellite. This action despins the satellite down to 10 rpm within a half-second. At this time the satellite is in a mission mode of operation. After the satellite has been in orbit for four or five months, the earth's magnetic field acting as an eddy current brake upon the satellite's ferromagnetic material, reduces the satellite's spin rate. When the spin rate falls below 8 rpm, a pair of spin-up rockets underneath the satellite's baseplate are fired. These small solid rockets burn for about a half-second and develop enough thrust to increase the spin of the satellite by an additional 3.5 rpm, returning it to mission mode speed again (nominally between 8 to 12 rpm). The first eight satellites carried ten of these rockets, whereas TIROS IX introduced a new system which maintained spin control by means of magnetic control torque.

The earlier TIROS TV system consisted of two redundant subsystems, each comprising a TV camera, tape recorder, video transmitter, and associated electronics. The TV system had a capability of storing 32 images on each video recorder and sequencing each picture at approximately $\frac{1}{2}$-min intervals. The TIROS IX satellite had a more flexible programmer that was capable of varying the picture interval from 32, 64, or 128 sec. The earlier TIROS satellites (II, III, IV, and VII) had a nonscanning wide-angle radiometer experiment, as well as an infrared sensor which had 5-channel scanning radiometers measuring from 0.5 to 50μ.

The next segment of Figure 4 shows the telemetry subsystems including the 136-MHz beacon, the 235-MHz 2-W TV transmitter, which was later changed to a 5-W solid-sate transmitter, and the necessary telemetry commutators to sample the various housekeeping functions aboard the satellite.

The last segment of Figure 4 shows the function control system, consisting of programming devices which automatically sequenced the satellite's cameras and tape recorders on the remote orbits when the satellite was beyond communication range of the two TIROS CDA ground stations. The CDA stations can send commands to the satellite programmers which, in turn, will sequence cameras and recorders to operate over the portions of the sunlit phases of the orbit to provide the desired overlapping picture swath. Power and weight are minimized by having these units operate only when they are required. To increase the mission lifetime, TIROS IX and ESSA were capable of cross-coupling the various subsystem chains; instead of operating either redundant System 1 or System 2, one could operate a segment of System 1 with segments of System 2 to provide additional reliability.

The camera system picture-transmission sequence from a TIROS satellite operates as follows. The wide-angle (104° lens) TV camera contains a half-inch vidicon with a frame rate of approximately 2 sec for its 500-line scan. The earth scene is exposed for $1\frac{1}{2}$ msec by the camera's magnetically operated focal-plane shutter. The picture is then read out electronically for a 2-sec period and is transmitted in real time in the direct mode of operation to the TIROS CDA station that is in line of sight with the satellite. During remote operation, the video data is stored on the video recorder, which is turned on for each picture sequence, for later transmission. When commanded, the stored pictures are transmitted, utilizing the 5-W TV transmitter to the receiving antenna at the CDA station. The video image is processed and read out on a 5-in. kinescope which, in turn, is photographed by a 35-mm camera. A polaroid camera can also be used. In this manner, earth-cloud pictures of a direct or remote orbit are received and processed.

The wide-angle, medium-angle, and the narrow-angle cameras employed on the earlier TIROS satellites are shown in Figure 5. The TOS system

Figure 5   Cameras employed on early TIROS satellites.

utilizes the advanced 1-in. vidicon cameras instead of the 0.5-in. vidicon cameras; also, a line scan rate of 800 rather than the 500-line system is employed in the APT and AVCS cameras used with the TOS system.

The TIROS CDA (Command and Data Acquisition) stations are located at Fairbanks, Alaska, and Wallops Island, Virginia. Each CDA station has the capability of command control of the satellite and can receive the satellite's TV and telemetry data. The station contains backup equipment so that in the event a component is inoperative, it has the capability of switching in redundant sections. It is important that the ground stations have redundancy to minimize the loss of vital data, since with the operational mission, as in the TOS program, the goal is to provide daily global coverage without interruption and loss of data.

Figure 6 is the world's first satellite TV picture brought back from space by TIROS I on 1 April 1960. This satellite demonstrated the feasibility of an earth-orbiting satellite's ability to bring back useful pictures of the earth's cloud cover by means of a slow-scan miniaturized television system. The first space TV photo (taken by a wide-angle TV camera system) shows a large frontal system, south of Nova Scotia. The scene covers about a million and a half square miles. This photo was a breakthrough for the weather satellite program; for TIROS I achieved its primary mission objective on its first orbit with the first picture.

Another significant discovery on TIROS I is depicted on the TV picture shown in Figure 7, which was also received on the first day. This cloud system is not the random scattering of clouds seen from the ground, but

Figure 6   TIROS I, orbit 1, frame 1. First television picture of earth's weather received from a satellite; scene shows weather front over Eastern U.S. and Canada.

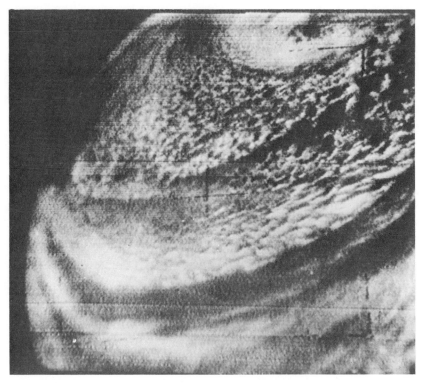

Figure 7   Weather on the grand scale, as photographed by TIROS I on its first day of operation.

is a highly organized global storm system. The storm center is near Alaska, but 1000 to 1500 mi south of the storm center the clouds are still well organized. One of the reasons why many meteorologists prefer the wide-angle system is because it affords a means to view the global weather systems shown above.

Figure 8 shows a narrow-angle picture taken by the TIROS II camera. Whereas the wide-angle pictures observed a scene that measured 800 by 800 mi when the cameras pointed along the local vertical, the narrow-angle camera photo viewed a scene 70 mi on a side. Figure 8 depicts the ice floes in the St. Lawrence River, Canada. This type of picture could provide information of sea ice conditions in the cold northern bodies of water.

TIROS III was launched in July 1961 to become the first satellite put on station to observe hurricanes for the U.S. meteorological program. Figure 9 shows a busy day for TIROS III on 19 September 1961. During

Figure 8   Ice floes in the St. Lawrence River, as photographed by TIROS II.

each orbit a picture swath approximately 5000 mi in length and 800 mi in width was taken during the 32-picture remote sequence.

The data obtained from the TIROS and ESSA satellites are transmitted to the U.S. Weather Bureau, where the information is analyzed and transformed into nephanalysis for facsimile transmission. The nephanalysis of the satellite TV data is then placed on photo-facsimile machines at the National Environmental Satellite Center (NESC), Suitland, Maryland, and transmitted to the major weather bureau centers in the United States. The meteorological team at the CDA stations formerly interpreted the pictures and issued "quick-look" data; however, all data since 1965 have been sent directly from the CDA stations to ESSA at Suitland, Maryland, by telephone line. At NESC, the data are analyzed and incorporated with other ESSA weather data at the National Meteorological Center. (In April 1962, the U.S. Weather Bureau instituted a system to broadcast the satellite weather data on an international basis in coded form to provide advance warnings of impending storms to nations of the world.)

Figure 10 shows some typical hurricane and typhoon cloud patterns. The very first hurricane detected by TIROS III was Hurricane Anna passing northwest of Venezuela's Lake Maracaibo. One of the figures shows Typhoon Ruth with a very prominent eye.

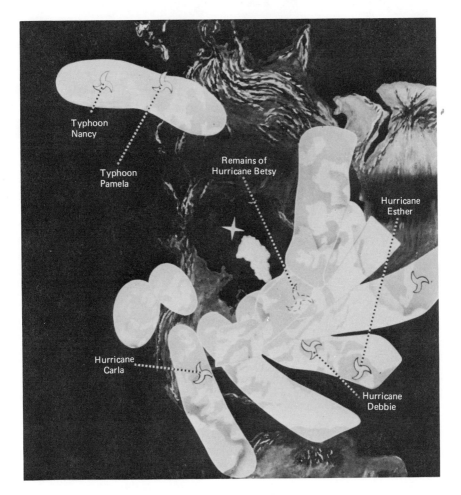

Figure 9   Weather activity observed by TIROS III on September 19, 1961.

TIROS IV was used in an experiment called Project TIREC (TIROS Ice Reconnaissance Program), which coordinated the TIROS satellite television data and pictures taken by four Canadian CF100 jets over the Gulf of St. Lawrence and the Gaspé Peninsula in Canada. The TIROS IV picture shown in Figure 11 was a single direct transmission showing

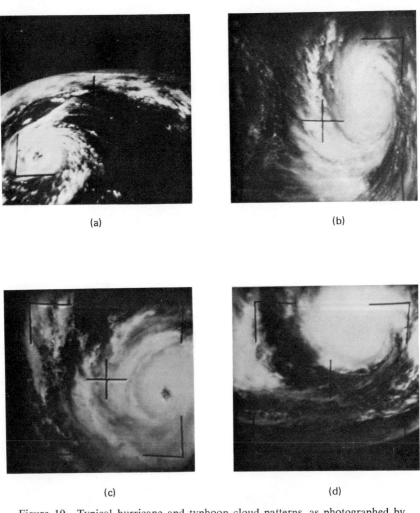

(a)   (b)

(c)   (d)

Figure 10 Typical hurricane and typhoon cloud patterns, as photographed by TIROS satellites. (a) TIROS VII orbit 1941: hurricane Ginny (1963). (b) TIROS VII orbit 6952: Hurricane Hilda (1964). (c) TIROS VII orbit 6746: Hurricane Gladys (1964). (d) TIROS VII orbit 10,745: Typhoon Dinah (1965).

Figure 11 Sea-ice surrounding Prince Edward Island in the Gulf of St. Lawrence, as photographed by TIROS IV for Project TIREC.

Prince Edward Island completely surrounded by sea ice, whereas the rest of the Gulf is free of ice. The aircraft had to make numerous flights which took many hours to cover the same area and had to utilize several hundred overlays to depict the test area.

Figure 12 shows a clear day over the eastern Mediterranean Sea showing Cyprus and the coast line of Israel and Egypt. Note the Nile Delta and the Suez Canal, and the sun glint evident on the Mediterranean. Hopefully sun-glint observations in the photos will lead to the determination of the sea state. Note also the darker surroundings indicating the presence of vegetation in Israel. The ability to determine the presence or the lack

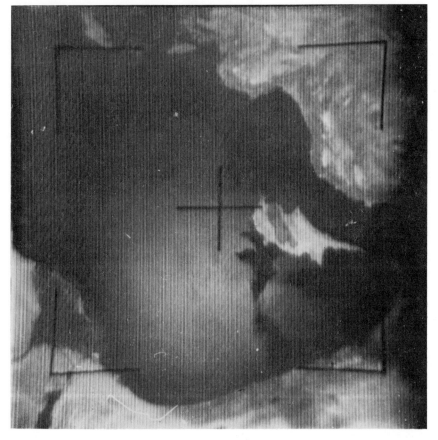

Figure 12   Cyprus and sun glint off Mediterranean Sea, as photographed by TIROS VII.

of vegetation by satellite-provided TV pictures will be one of the objectives of the future Earth Resources Satellites.

Figure 13 shows a TV photograph obtained from TIROS VII during the 1965 hurricane season.

The Automatic Picture Transmission (APT) was first used on TIROS VIII and was unique insofar as it permitted nations all over the world to receive the satellite data in real time. The camera has the capability of photographing the local scene by means of a special 1-in. vidicon camera that has a very slow readout time. Instead of a 2-sec readout as on the 0.5-in. vidicon TIROS cameras, and 6-sec readout on Nimbus and ESSA 3 AVCS cameras, the APT camera has a 200-sec readout.

With this slow data transmission rate, a very simple ground station capable of accepting data at 2-kHz bandwidth can be used. A one-man station can record the APT camera information directly by means of a facsimile printer, instead of going through photographic processing. In a matter of 200 sec as the satellite transmits the data, the pictures are back on earth and ready for analytical use. During the TIROS VIII experiment, there were 40 such sites located around the world. With the TOS ESSA 2 satellite system there are more than 170 stations around the world receiving APT pictures directly.

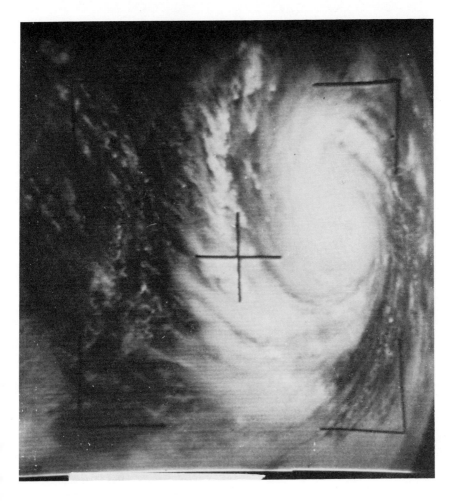

Figure 13   A TIROS VII photograph.

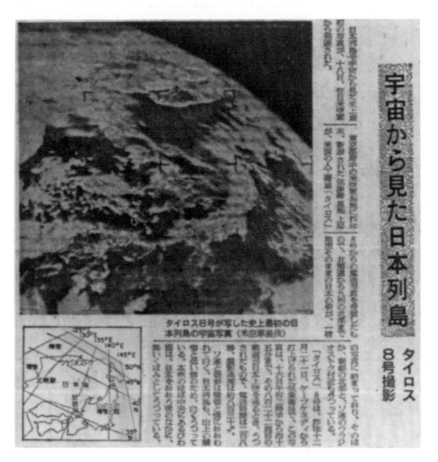

Figure 14   A TIROS VIII APT photograph as received in Tokyo, Japan, and reprinted in local newspapers.

Figure 14 is one of the pictures of the islands of Japan transmitting by TIROS VIII and received in Tokyo. The Japanese government released this photo for their newspapers. The copy shown here is approximately the tenth-generation copy.

The TIROS wheel configuration has dimensions similar to the previous TIROS satellites—42 in. in diameter by 22 in. in height. In TIROS IX the two TV cameras were canted so that contiguous global coverage could be achieved from an orbital altitude of approximately 400 nmi. With the TOS configuration, the cameras are mounted parallel to the baseplate instead of being canted. At the TOS mission altitude of —750 mi, one APT or AVCS camera viewing the earth on a daily basis.

Figure 15 is a view of the TIROS IX baseplate with its solar array removed; as can be seen in this illustration, the video recorders and the two canted cameras are diametrically mounted. As the satellite spins in orbit, with the spin axis normal to the orbital plane, each camera has an opportunity to look straight down and, hence, effectively take pictures along the local vertical, thereby achieving the same results as a fixed earthoriented platform.

Figure 16 shows the sensor layout, the cameras, and its infrared (IR) sensors. As the satellite spins at approximately 10 rpm, the IR sensor intercepts the leading edge of the earth; after scanning the cold body of space, the sensor detects the warm body of earth and activates the camera-triggering circuits at the time the camera is pointing along the local vertical. Taking a picture once a revolution would result in wasteful overlapping at the 10-rpm spin rate; therefore, the TV camera is programmed to take a photograph once every 2 min on TIROS IX, every 6 min on ESSA 2, and every 4 min on ESSA 3, which is adequate for the required picture overlap. The satellite has dual IR sensors to trigger each camera, which provides the necessary redundancy. In addition, these same IR sensors are used to commutate the spin control coil and effectively control the satellite's spin speed by means of the MASC (Magnetic Attitude Spin Control) mode of operation.

On TIROS I the spin axis precessed approximately 3° to 4° per day After evaluating all the perturbing torques, Dr. W. Manger of Radio Corporation of America developed a mathematical model which indicated that the earth's magnetic field was the primary force. This force, acting upon the satellite's residual magnetism, caused the spin axis to precess. This information was applied on TIROS II, on which the spin axis precession was effectively controlled. By placing a coil of wire around the periphery of the satellite and by controlling the amount of current flowing in the coil, a controllable magnetic field is induced which interacts with that of the earth's magnetic field in space and results in a controlled steering torque for maneuvering the satellite's spin axis, thus providing a means of effective attitude control. With this scheme it was possible to optimize the photo-coverage of the earth and the thermal control of the satellite. This technique was used very effectively on TIROS II through TIROS X and then on the ESSA satellites. On TIROS IX and on the ESSA satellites the techniques of magnetic torquing were refined to obtain more effective control and greater precision and flexibility in attitude positioning of the satellites.

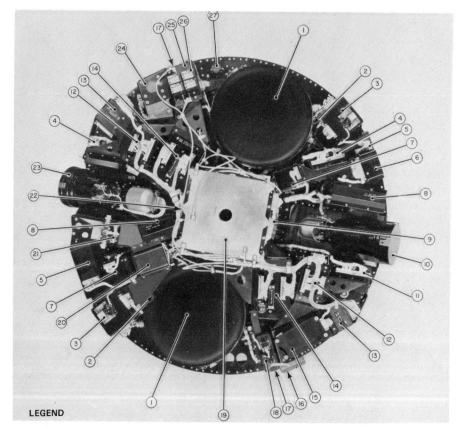

LEGEND

1. Tape Recorder
2. Power Control Unit (Hidden)
3. Recorder Electronics
4. Command Control
5. Recorder Power Converter
6. MBC Switch
7. Voltage Regulator
8. Camera Electronics
9. Command Distribution Unit
10. TV Camera No. 2 (Viewing 26.5° Upward)
11. Auxiliary Control
12. Dyson
13. Computing Trigger
14. Digital Clock
15. Power Supply Protection Unit (Hidden)
16. Solar Aspect Indicator (With Cover)
17. Two Orthogonal Horizon Sensors (Hidden)
18. Solar Aspect Indicator Electronics
19. Antenna Coupling Network
20. TV Transmitters and Video Coupling Network
21. Command Receivers
22. Battery Pack (Hidden)
23. TV Camera No. 1 (Viewing 26.5° Downward)
24. Telemetry Commutators and Signal Conditioner
25. Beacon Transmitters
26. Subcarrier Oscillators
27. V-Head Attitude Sensor

Figure 15   TIROS IX equipment layout.

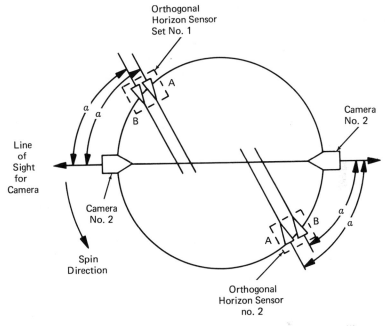

Figure 16   Sensor layout on the baseplate of a TIROS wheel satellite.

Figure 17 shows the two magnetic coils that were employed on TIROS IX. The first is a coil which is concentric to the baseplate. This coil creates a dipole moment that is in line with the spin axis. By means of this QOMAC (Quarter Orbit Magnetic Attitude Control) coil, the satellite can be maneuvered from its initial injection position in orbit where the spin axis initially is in the orbital plane to the wheel or mission mode, wherein the spin axis is maneuvered so that it is normal to the orbital plane. This coil has two modes of operation: a high-torque mode and a low-torque mode. The high-torque mode is employed during the turn-around maneuver, permitting the movement of the satellite's spin axis of approximately 10° per orbit. Once the mission mode is achieved, the low-torque coil can be used to maneuver the satellite about 2° per orbit. This coil needs to be activated no more than once every 5 or 6 days on TIROS IX to keep the satellite within 1° of its nominal attitude despite the highly elliptical orbit. In fact, on ESSA 2 and ESSA 3 the attitude control coil is used about twice a month to hold this accuracy continuously. This system is not as time-limited as other control systems which require stored gas; also, because of the low duty cycle of the coil,

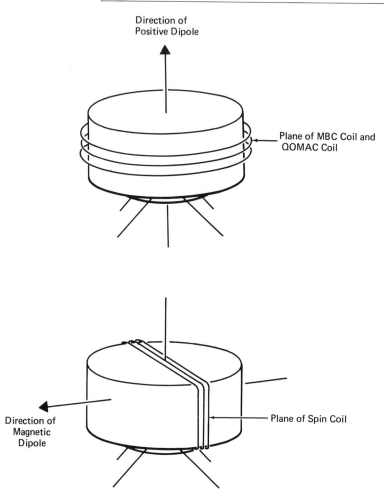

Figure 17   Magnetic coils employed on a TIROS wheel satellite.

the average power used for the magnetic attitude control is on the order of milliwatts.

Since most satellites have a residual magnetic field and since some of the materials used aboard the satellite could change magnetic characteristics (such as the nickel-cadmium batteries), a MBC (Magnetic Bias Control) Coil is provided on the satellite. This coil can be controlled from the ground so that a selected magnetic bias can be effected to compensate or null out any on-board residual magnetism. The residual magnetic effects on the satellite are determined by taking data over some period of time.

The MASC (Magnetic Spin Control) Coil was added for the first time on TIROS IX. The coil lies in the plane of the spin axis and provides a dipole moment normal to the spin axis. By commutating this coil for short periods of time, one can cause the satellite to act as a motor in space and either spin-up or spin-down the satellite's spin rate. The commutation of the coil is effected by the two infrared scanners which change the coil polarity every half spin. By this technique one can maintain precise spin control to 25 msec. This is the primary mode of operation employed on TOS.

Instead of having the satellite vary between 8 and 12 rpm, TOS operates on a precise spin rate, and the satellite spin rate is very effectively controlled by the MASC coil to provide precision spin control so that the spin of the satellite can be used to synchronize the vertical sweep of the cameras and be used as a counter for picture-interval sequencing. The QOMAC, MBC, and MASC coils are reliable, since there are no moving parts other than a stepping switch which changes current values.

Figure 18 shows the V-head scanner used to determine the attitude of the satellite. Again, a very effective technique is employed with a simple device to derive precise attitude data. These sensors "look out" covering a 100° angle with the bisector normal to the satellite's spin axis. If the satellite is on track, the two sensors "sweep out" arcs from the earth's leading horizon to the trailing edge of the horizon which are equal in time duration. If the satellite spin axis is displaced in attitude, the sensors sweep out unequal arcs. The data yielded by the V-head scanner is received at the CDA station during a 10-min contact via the telemetry transmitter. By manual or computer methods, the differences between arcs are measured to determine the satellite's attitude error. The error can be determined within 0.1°. If the error exceeds the nominal desired 1.0° value, a corrective signal is sent to the satellite's attitude control coils to restore the spin axis to nominal position. TIROS IX was inadvertently placed into a highly elliptical orbit with an apogee four times higher than expected, and the sensors and the control system functioned very satisfactorily in spite of the large variations of the orbital parameters from the nominal planned mission, again demonstrating the flexibility of the system.

Figure 19 shows a sequence taken by TIROS IX, camera 1 and camera 2 being triggered within half a spin of each other, the sequence starting from the South Pole going northward along the coastline of South America. One can see the effective stability of the satellite's attitude system. These pictures are taken at about 1-min intervals.

In the TIROS IX photo display presented in Figure 20 a series of

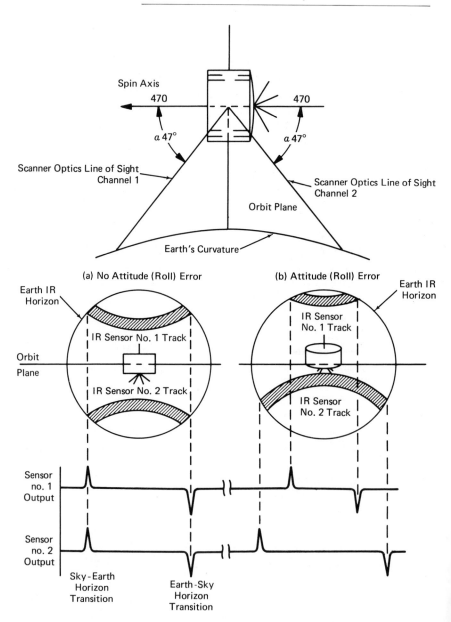

Figure 18   Output of V-head scanner employed on TIROS wheel satellite.

End of
Sequence

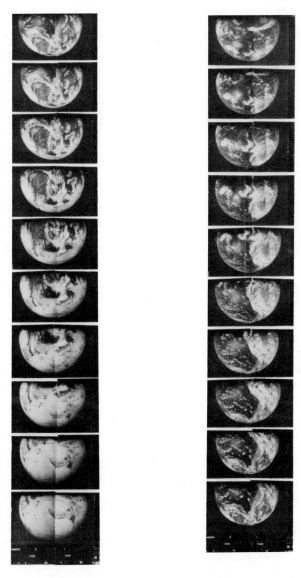

Start of
Sequence

Figure 19   Sequence of pictures taken by TIROS IX on orbit 115: sequence
starts at lower left with pictures of antarctic ice cap and ends at upper right
with pictures of central America.

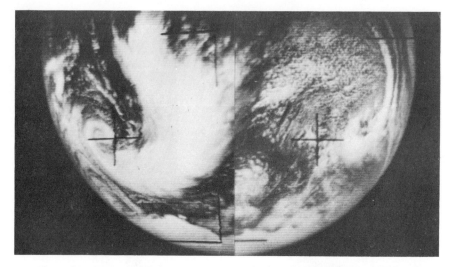

Figure 20   Large storm vortex near Antarctica, as photographed by TIROS IX (orbit 436).

massive storms can be seen. The picture resolution is four times greater than expected owing to the altitude anomaly, and it is evident that the data on this pair of pictures is extremely useful. This storm is approximately 2000 mi in dimension. The ice cap of Antarctica can be seen in the lower part of the photo.

Figure 21 shows a sequence taken by TIROS IX demonstrating the unique ability for this system to observe the entire earth in one day of satellite operation (a total of 450 photographs taken in over 12.5 orbits).

Figure 21   First complete view of the world's weather. This photomosaic is composed of 450 photographs taken by TIROS IX during its 12 orbits on February 13, 1965. (Courtesy of U.S. Weather Bureau)

A number of storms can be noted in this sequence. This was the major goal in the TIROS wheel mission—to demonstrate the daily global observation capability of an orbiting satellite. TOS is capable of providing global daily coverage, but because of its 750-nmi altitude and 1-in. rather than 0.5-in. vidicon TV camera, only one camera is required. The second camera is provided strictly as a back-up system.

Table 2 shows the performance of TIROS IX in the first five months or orbital operation. The countries listed received warnings of major storm systems; some were warned up to four or five days in advance. In six months of operation of one satellite, the U.S. Weather Bureau published 295 storm warnings to countries other than the United States.

Table 3 presents a summary of the useful life in orbit and the number of TV pictures returned in that time for the 10 TIROS and 5 ESSA satellites placed into orbit over the past seven years. All launchings were suc-

TABLE 2
COUNTRIES AND WORLD AREAS FOR WHICH STORM
ADVISORIES WERE ISSUED FROM TIROS IX DATA DURING
THE SATELLITE'S FIRST FIVE MONTHS OF OPERATION

Australia	41
Burma	1
Ceylon	5
China	1
East Africa	13
Fiji	8
Guam	1
Bombay	36
India	16
Indonesia	7
Malagasy	26
Malaysia	5
Mauritius	46
New Caledonia	2
New Zealand	15
Pakistan, East	10
Pakistan, West	3
Philippine Islands	1
Portugese East Africa	4
South Africa	43
South Viet Nam	3
Tahiti	5
Thailand	3
Total	295

TABLE 3
TIROS/TOS PERFORMANCE SUMMARY

Satellite	Useful Life (Days)	Total TV Pictures
TIROS I 4-1-60	89	23,000
TIROS II 11-23-60	376	36,100
TIROS III 7-12-61	230	35,000
TIROS IV 2-8-62	161	32,600
TIROS V 6-19-62	321	58,200
TIROS VI 9-18-62	389	66,600
TIROS VII 6-19-63	1406*	124,440
TIROS VIII 12-21-63	1221*	102,480
TIROS IX 1-22-65	823*	86,310
TIROS X 7-1-65	663*	76,600
ESSA 1 2-3-66	446*	107,570
ESSA 2 2-28-66	421*	37,900
ESSA 3 10-2-66	205*	30,500
ESSA 4 1-26-67	89*	8400
ESSA 5 4-20-67	4*	40
Totals	6844	825,740

* Still useful as of 24 April 1967.

cessful; and with each satellite, the design goal for life in orbit was exceeded. The last of these satellites are still functioning; however, the ESSA satellites are primarily used for the daily operational global forecasts whereas the other satellites are for back-up use. More than 825,000 pictures have been transmitted to date by these 15 spacecraft.

The TOS system consists basically of two configurations, as shown on Figure 22: one carrying APT cameras for direct local readout to simple stations located anywhere around the world, and a second system carrying AVCS (Advance Vidicon Camera System) cameras for global readout at NESC.

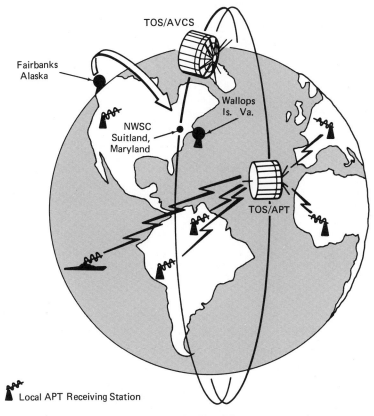

Figure 22   The basic TOS system.

The APT readout station is a one-man station which consists of a facsimile recorder with necessary receiving gear and a steerable and remotely slaved antenna which tracks a signal received from the satellite. An antenna beam width of 30° will permit the reception of the APT pictures. As noted, the picture images are read out over 200-sec periods, so that in one orbital pass there is time to read out approximately three to four APT photographs. Each APT station can receive the satellite data on two or three orbits a day; this means that each APT station

Figure 23   Photo-coverage capability from a TOS satellite.

can receive between six to eight photographs a day. The daily orbital sequence received by a single APT station will provide a scene 2000 mi in radius, enough to forecast for a 24-hour period.

Figure 23 depicts the APT or the AVCS photo-coverage capability from the TOS satellite. One camera at the 750-nmi altitude will provide a scene 1700 by 1700 nmi, or approximately 3 million square miles per picture. The TOS AVCS system, which provides the remote storage read-out capability, will take 12 pictures per orbit with 50 percent overlap between frames; whereas the APT system will take 8 pictures per orbit with approximately 30 percent overlap between frames. Both systems provide contiguous coverage at the equator between adjacent orbit sequences and hence, daily global coverage. Figures 24 and 25, respectively, are

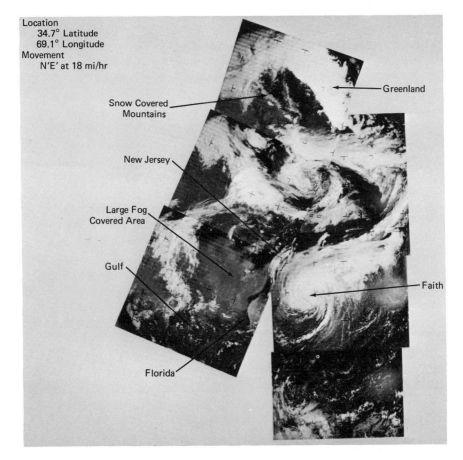

Figure 24 Hurricane Faith, as photographed by Essa 2 on September 1, 1966.

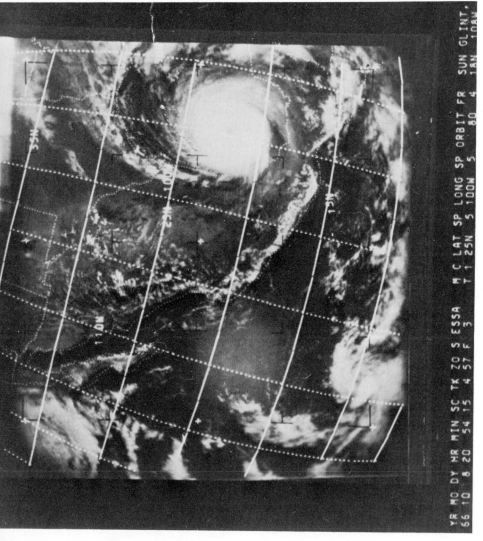

YR MO DY HR MIN SC TK ZO S ESSA   M C LAT SP LONG SP ORBIT FR SUN GLINT,
66 10 8 20 54 15 4 57 F 3   T 1 25N 5 100W 5   80 4 1.8N 108W

Figure 25   ESSA 3 television photograph of Hurricane Inez over Mexico's Yucatan Peninsula (gridding and identification data added by digital computer at the National Environmental Satellite Center of the Weather Bureau).

representative of the TOS APT and AVCS photographs obtained from ESSA satellites.

The devices and techniques employed on TOS are similar to those on TIROX IX, except that they utilize improved circuits and advanced cameras, and the subsystems were designed to withstand the more severe radiation environment at the TOS orbital altitudes. The TOS satellite uses integrated circuits and a 1-in. vidicon camera system which is fully redundant with cross-strapping of many units. For example, if Side 1

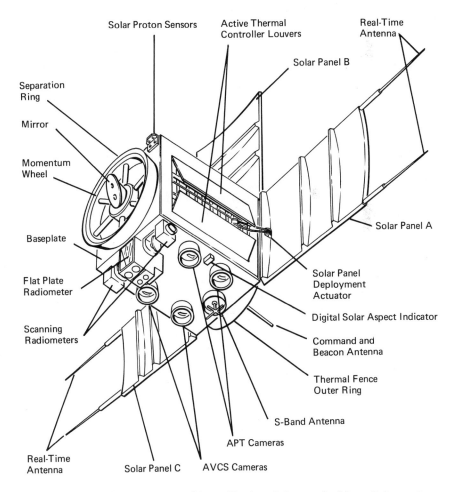

Figure 26   The TIROS MITOS satellite in mission mode. Note: Solar proton sensors are shown here for clarity. Actual location is in corner concealed by momentum wheel.

transmitter fails, Side 2 transmitter will transmit Camera 1 data, or if Camera 1 fails, the balance of the equipment can be used with Side 2.

Like TIROS IX, the TOS satellites are configured with the QOMAC and MASC magnetic coils; in addition, a liquid damper was added to ESSA 2. All satellites up to ESSA 1 used a mechanical precision damper capable of limiting nutation to 0.5°; however, a liquid damper device which limits nutation to 0.1° was added on ESSA 2 and used on ESSA 3, 4, and 5.

The system beyond the present TOS satellites will be an infrared day-and-night observation satellite. For this case a wheel satellite could be used to carry an IR device which would be capable of mapping out the earth both day and night, where the present system is limited to daytime operation. A wheel system could also be used with phase-locking of the IR scanning with the satellite spin rate.

An alternate and more probable approach is the use of a stabilized platform employing a stabilite control system as shown in Figure 26. An evolution from the spin-stabilized satellite to the stabilite configuration is now under design for the second-generation TOS mission. We have gone from a standard TIROS I with spin control only, to TIROS II through TIROS VIII with limited magnetic torquing. In TIROS IX and X and ESSA 1 through 5, refined magnetic torquing control was employed. Stabilite is being plannned for the next-generation TOS satellites, which require a despun platform because of the variety of sensors required for future missions and the conflicting requirements of these sensors were a spinning platform to be used. To maintain the same simplicity, reliability, and accuracy of attitude control achieved in the previous TIROS and TOS satellites, the stabilite technique and magnetic torquing, which are related, were chosen for the three-axis attitude control of the sensor platform.

STANLEY WEILAND

Stanley Weiland studied physics at Queens College and received his B.A. degree in 1948. He is currently the Nimbus Observatory Systems Manager for the Goddard Space Flight Center, Greenbelt, Maryland. Mr. Weiland directs the spacecraft systems design and is thus in charge of the interface between the satellite and the many-changing meteorological experiments which must be fitted into it.

Abstract—*The Nimbus is a second-generation experimental weather satellite. It is larger than TIROS and is designed to be an orbiting laboratory for the development of the sensors which are ultimately to collect the data for the global weather system. Mr. Weiland describes the Nimbus system; its launch procedure, power supply, attitude control systems, and finally its daylight and infrared cameras. Included are a number of cloud-cover pictures taken by Nimbus I and II.*

The Nimbus Satellite System

THE Nimbus spacecraft system represents a major step forward from the TIROS weather satellite, both in the technological and scientific areas. To give some idea of the comparison, Nimbus is roughly three times as heavy as the TIROS or the TOS (TIROS Operational Spacecraft) spacecraft. It develops about three times the power and it can carry about three times the sensory payload of TIROS. The Nimbus spacecraft is earth oriented, nominally a 1000-lb, fairly high powered meteorological observatory spacecraft.

In this context, the word "observatory" means flexibility; it means that Nimbus is adaptable from one launch to the next, so that it can proceed with the general mission of meteorological research and development from a satellite and can accept new experiments on each mission.

The Nimbus spacecraft is injected into a 600-nmi sun-synchronous orbit inclined about 10° off the pole in a retrograde orbit. Figure 1 demonstrates the orbital path of the Nimbus spacecraft and the type of coverage that a high-noon sun-synchronous retrograde orbit can give. High-noon sun synchronous means that the orbit plane contains the earth-sun line. As a result, it is possible to maintain optimum acceptance of solar energy throughout the orbit by rotating the paddles about an axis orthogonal to the orbit plane. The first advantage of this system very obviously is the ability to collect a great deal of solar energy and, therefore, to develop usable power to operate the spacecraft. The second advantage in the sun-synchronous orbit is that it gives continuous and complete earth coverage with the high-quality AVCS (Advanced Vidicon Camera System) optical sensors. This presumes, of course, a proper design of the optical system.

At this time, the Nimbus spacecraft program consists of three flight-approved spacecraft: the Nimbus I, which was launched in August 1964; the Nimbus II, the second in the series which was launched in May 1966 from Vandenberg; and the Nimbus B, the third in the series which was launched in the last quarter of 1967. The fourth Nimbus spacecraft, the so-called Nimbus D proceeded to the flight hardware stage in mid 1966,

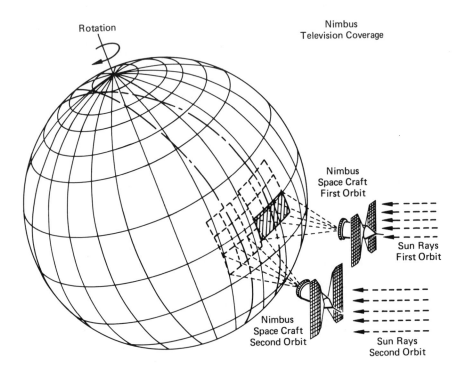

Figure 1   Orbital path of Nimbus and TV coverage.

and to a launch date in late 1969. Thus Nimbus is a continuing research-and-development meteorolgical satellite program.

Figure 2 shows the components of the spacecraft. The Nimbus space-craft consists of three major elements: the first is the so-called sensory ring, which is shown at the bottom of the figure. This is a torus-shaped structure with a center section adaptable to each particular launch. The shape of this inner structure is tailored to the experiments that are to be flown. In this case, we have the Nimbus I configuration launched in 1964 with the three high-resolution cameras, the tape recorder for the cameras, the S-band antenna, the high-resolution radiometer, and the slow-scan APT (Automatic Picture Transmission) camera. This is a clearly identifiable section of the spacecraft, and it is handled as a unit during construction. The sensory ring is assembled and a number of tests are performed on the ring before its integration into the full spacecraft configuration. The second clearly identifiable section of the spacecraft is the two solar paddles which together represent about 50 ft² of solar cells. The third section and a very significant one, is the attitude control system, which is the boxlike structure on the top of the spacecraft. The spherical shape beneath the attitude-control section is the cold-gas bottle which contains the Freon-14 gas for the control jets. The structure is interconnected by way of the truss assembly, which also serves as a continuation of the ground plane for the command antenna on the top of the spacecraft.

Figure 3 gives a more detailed view of the attitude control system of the spacecraft. It is a three-axis stable active control system which employs, in pitch and roll, horizon infrared horizon scanners. The outputs of the scanners are processed in a digital attitude computer, and if there are attitude errors in either pitch or roll, these are sent as signals to a reaction wheel. There are three of these wheels, one in each of the

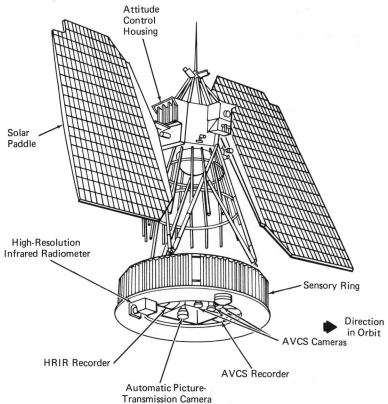

Figure 2   Components of Nimbus 1.

three axes, which store by direction any momentum which may be imparted to the spacecraft. Torques may be caused by the action of the solar paddles which are driving in the pitch direction, or they may be caused by solar pressure, aerodynamic drag, or by magnetic torques. When the wheels can no longer absorb the momentum, a short pulse of gas is fired to provide an opposite force to bring the spacecraft back within the control of the wheels. A gyrocompass is employed as a sensor in the yaw axis. On the right of the figure are the forward scanner and the albedo shield, and to the left, the shaft upon which the solar paddles are mounted (the solar-array drive mechanism being a portion of the attitude control system).

Also shown on the left is an element of the active thermal-control system. This is a louvered shuttered assembly which is open or closed, depending on the temperature in this particular location (a similar one is on the other side). Opening the shutters allows more energy to be radiated into space and hence, cools the spacecraft. There are also passive thermal controllers in the form of insulation and thermal paint. The com-

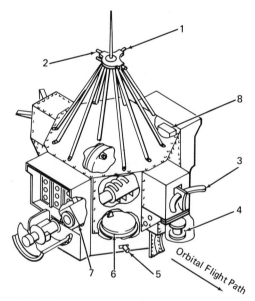

1.	Roll Nozzles (1)	5.	Yaw Nozzles (4)
2.	Pitch Nozzles (2)	6.	Flywheels (3)
3.	IR Horizon Scanners (2)	7.	Solar Array Drive Mechanism
4.	Coarse Sun Sensors (2)	8.	Gyro

Figure 3   Attitude control system.

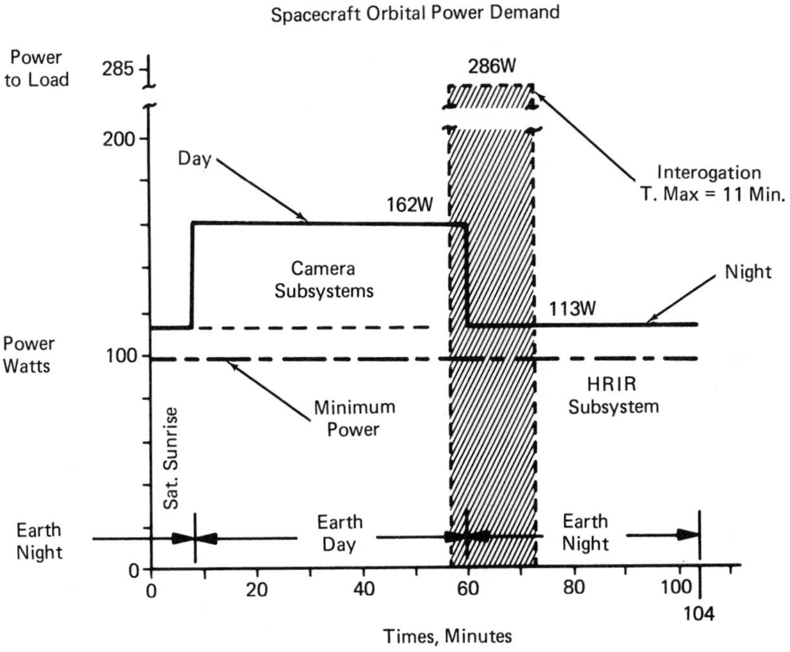

Spacecraft Orbital Power Demand

Figure 4   Spacecraft orbital power demand.

bined active and passive control systems allow the spacecraft interior to remain at $25° \pm 10°$ C.

During the daytime portion of each orbit, the spacecraft develops about 450 W of unprocessed power from the solar paddles. This solar energy is used for two purposes. First, the output is used directly to power the spacecraft during the daytime portion of the orbit. Second, the excess energy not required to power the spacecraft electronics is used to charge nickel cadmium batteries, which deliver useful energy to maintain operation during the nighttime portion when the spacecraft is in the earth's shadow. Additionally, once and sometimes twice an orbit, when the spacecraft passes over the command and data-acquisition ground station, the command is given for the spacecraft to transmit all the stored data. During this period, the power demands increase considerably because the high-powered transmitters on the spacecraft are operating.

Figure 4 shows the type of duty cycles imposed upon the power sources. The orbital period shown is 104 min, which is a nominal value for 600 mi. Since the satellite is orbiting at 600 mi, we make a distinction between satellite day and earth day because the satellite's period in the umbra

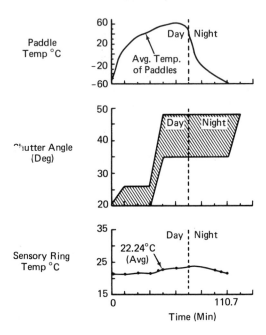

Figure 5 Nimbus thermal control subsystem.

is somewhat shorter than the earth's night. The minimum power demands of the basic spacecraft, the attitude control system, the telemetry system, and the command system are shown by the lower dashed line. The operation in the daytime is shown with the cameras operating (which have a requirement of 162 W). At the time of acquisition, shown by the cross-hatched section, the power demands rise very sharply during transmission of the data to the ground. This period, as shown, overlaps a day-to-night transition between earth day and earth night, which is quite common for the acquisition ground station at Fairbanks, Alaska. Finally, the night-time power demands are shown to be 113 W.

Figure 5 gives some indication of the thermal behavior of the spacecraft. The top curve represents the thermal behavior of the solar paddles, which are exposed alternately to the sun's radiation and to cold space. As a result, the variations in temperature of the solar paddles are quite high. As can be seen, there is a range of $-60°$ to $+60°$C, but this is well within the design criteria established for the paddles, which is $+80°$ to $-80°$C.

The sensory ring shown on the bottom curve, on the other hand, must provide a benign temperature environment for several reasons. First, it is essential for the reliability of the electronic hardware, and second, the characteristics of the Vidicon cameras used as the main optical sensors are very sensitive to temperature variations. As a result, it was essential to maintain good thermal control in the ring. Our specification was to operate at an average temperature of 25°C with a tolerance of ±10°. The actual performance was considerably better, as shown by this curve.

The Nimbus data system consists of two major elements: the spacecraft portion and the ground complex. In the spacecraft system, we use a pulse-mode modulation low-data-rate telemetry subsystem, which is used to monitor the various housekeeping functions of the basic spacecraft and the experiments. It is also used to handle digital data from experiments with moderately low-data rates. For handling video and high-rate digital data, we use a frequency-division multiplex system which is FM-FM, on an *S*-band transmitter link. Nimbus I and II carried a 5-W transmitter; Nimbus B contains a completely solid-state 2-W transmitter.

Figure 6 depicts the entire Nimbus system. On the left is the Nimbus

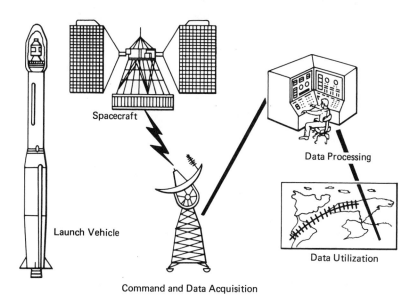

Figure 6 Nimbus system complex showing spacecraft, data processing, and ground links.

Satellite on top of the Agena launch vehicle. The 85-ft dish at Fairbanks, Alaska, is the primary data acquisition center. However, the demodulated receiver output is transmitted by land lines to the Goddard Space Flight Center for processing. The entire output fom one pass can be processed within one orbit period so there is no data stackup. The spacecraft command signals are sent in real time from Goddard via long lines to Alaska, and out the antenna. Thus, all the data are received and all commands originate in Maryland, even though the action is taken by way of the Alaska link.

The transmission of the video data is restricted over the long line between Fairbanks and Goddard. This is primarily a matter of logistics and economics, in that the multiplex output is roughly three-fourths of a megahertz and to set up a 1-MHz data link from Fairbanks, Alaska, to the Washington, D.C., area, would be extremely costly and difficult. As a result, the data are stored on magnetic tape at Alaska and is played back slowly at an 8:1 rate. (There is still sufficient time between passes to accept these data, process them, and be ready for the next spacecraft pass.)

Figure 7 compares the design goals with the actual performance of the Nimbus I. The design goals for the attitude control system as shown were recognized before launch to be goals and not specifications. The performance as shown on the right-hand chart is optimistic with one exception. We maintain pitch and roll within $2°$ and $3°$, respectively, but the yaw departed as much as $10°$ owing to error signals generated by cold clouds at the horizon. This is a very serious problem on some spacecraft, and on Nimbus, it turned out to generate some errors, but it occurred

NIMBUS 1 FLIGHT PERFORMANCE

ATTITUDE CONTROL	DESIGN OBJECTIVES	ACTUAL
PITCH	$\pm 1°$	$\pm 2°$
ROLL	$\pm 1°$	$\pm 3°$
YAW	$\pm 1°$	$\pm 8° - 10°$
PADDLE POINTING	$\pm 10°$	$\pm 2°$
POWER (MAX)	450 WATTS	470 WATTS
POWER (AVG)	170 WATTS	160 WATTS
CURRENT (AVG)	13 AMPS	13 AMPS
THERMAL CONTROL		
SENSORY RING (AVG)	$25° \pm 10°C$	$22° \pm 1\frac{1}{2}°C$
CONTROL HOUSING (AVG)	$25° \pm 10°C$	$25° \pm 3°C$
DATA TRANSMISSION	400 MILLION BITS/ORBIT	ACHIEVED

Figure 7   Nimbus 1 flight performance.

NIMBUS I

SENSOR PERFORMANCE

	DESIGN OBJECTIVES	ACTUAL
AVCS	RESOLUTION $\frac{1}{2}$ MI. GRAY SCALES 8	ACHIEVED
APT	RESOLUTION $1\frac{1}{2}$ MI. GRAY SCALES 6	ACHIEVED
HRIR	RESOLUTION 5 MI. TEMP RANGE 190°K TO 315°K	ACHIEVED

Figure 8   AVCS, APT, and HRIR sensors.

less than 5 percent of the time and was not critical to the actual operation of the attitude control system. The power supply and data system performed quite well, and the thermal control systems were excellent.

The three sensors flown on Nimbus I, as shown in Figure 8, were the AVCS (Advanced Vidicon Camera System), the APT (Automatic Picture Transmission), and the HRIR (High Resolution Infrared Radiometer). All three performed according to design goals, the AVCS maintaining a ground resolution of half a mile and a dynamic range of eight gray levels. The HRIR was originally intended to generate analog data, line by line, which would be digitized and computer-processed. A graphic display was added, in the form of a facsimile recorder. The APT system is a real-time readout camera. When the shutter opens, it exposes the optical image on the video, which is stored as charge on the photoconductor of the Vidicon and is read by the electron beam at a very slow rate, taking 200 sec to read out one frame. As a result, it is possible to transmit a high-quality, high-resolution picture slowly over a narrow communication link. Figure 9 shows the ground station equipment required to pick up the APT data. On the left is the helical antenna, its pedestal, and some connection cabling; on the right, the rack which involves a moderate-quality FM receiver of a police radio type, some demodulation electronics and a facsimile machine. About 60 production models of APT ground stations were built and were furnished to a wide variety of users: the Weather Bureau, the military, the Air Force, the Navy, and a number of foreign governments. As a result, APT stations exist all over the world today. In addition, the design is so simple that many individuals have built their own ground stations to handle the APT. Most of those built independently are of much better quality than the commercial stations. For instance, some of them use oscilloscopes as the facsimile machine

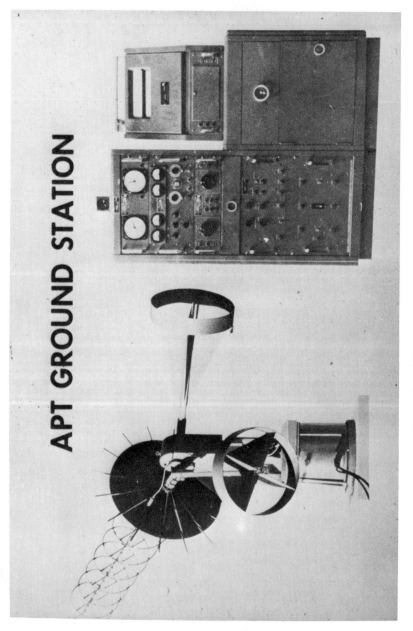

Figure 9   APT ground station equipment.

and simply photograph the oscilloscope image to produce the weather picture.

Figure 10 is a photograph of a cloud formation over Guadalupe Island which demonstrates the high resolution capability of the AVCS camera. It has a limiting resolution of about half a mile at an altitude, in this case, of approximately 300 nmi. Figure 11 is of no meteorological significance at all, but terrain features are always of interest. In this picture one may recognize the Suez Canal and the Nile Delta when it was the flood stage. It is an indication of the clarity and excellent performance of the Vidicon cameras.

Figure 12 shows the coverage of the APT (Nimbus I coverage), the cross in the center indicating the location of the ground station. On the previous orbit, the satellite, passing from south to north along a line through the center of the right-hand group of pictures, transmitted four photos to the ground station. On the next pass, the satellite transmitted four pictures and part of another to the station, and on the third pass, as it moved off to the left of the figure, the satellite transmitted only one picture. Once a day at noon in the United States, a local user can

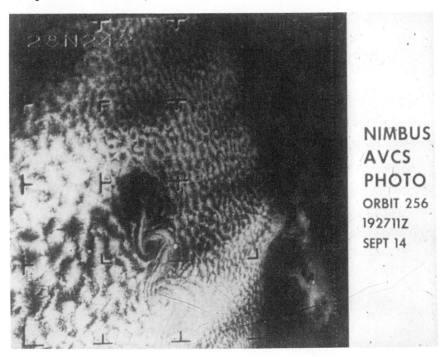

Figure 10   AVCS cloud cover photo over Guadalupe Island.

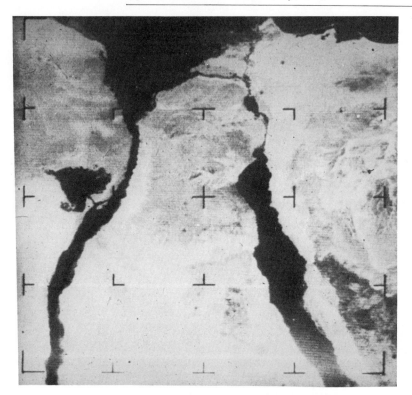

Figure 11   The Suez Canal, Egypt, and the Nile.

get a visual picture of the cloud cover over a large area surrounding his station. In general, it is over an area large enough so that once-a-day coverage will satisfy his needs.

Figure 13 is a series of single-pass pictures mounted together to give an idea of the extent of the coverage indicated by the center group of pictures in Figure 12. Notice that in the reproduced form, the quality of these real-time pictures is not comparable to the high-quality advanced Vidicon. However, the problem is basically in the ground presentation, rather than the camera. We have, for example, operated this camera in the stored mode with results indistinguishable from the quality of the AVCS pictures (for example, the picture of the Suez, Figure 11).

Figure 14 is a facsimile display of the HRIR which covers an extremely wide area, since the HRIR has a ground resolution of approximately 51 mi. Notice the boot of Italy and the Island of Sicily in the center. Notice also the gradations in the picture between black and white which point to an important difference between the radiometer and the camera.

This picture which was taken at night detects thermal radiation, the white areas representing the coldest temperatures, the black the warmest. As a result, there is a gradation, a certain Z-axis effect from the cloud formations, some of which are colder (the very high storm clouds) and some warmer (the low-level clouds).

Figure 15 is a more obvious example of this effect. It is an HRIR picture of the hurricane Gladys, one of the largest hurricanes ever photographed. Since the HRIR is a thermal detector, it is calibrated in temperature (degrees Kelvin here). The video shown represents the single analog trace of the temperature over the hurricane (plotted below the picture). The dark areas around the outside represent ground temperatures, the gray traces along the edges of the storm, clouds which are at moderately low elevations, and toward the center, the white areas represent the powerful hurricane clouds which are very high and, therefore, extremely cold,

APT Picture Coverage for One Day at One Station

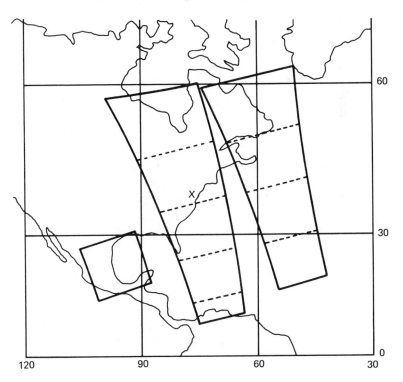

Figure 12   Nimbus 1 APT ground coverage for one day at a station along U.S. southeast coast.

with temperatures down to about 220°K. It is remarkable to note the dark spot in the center of the picture where the temperature rises very rapidly and corresponds to the instrument viewing through the eye of the hurricane and looking directly at ground temperature.

Recall that the chronology of the Nimbus missions A, C, B, and D is in slightly rearranged form. We will look briefly now to the Nimbus C, B, and D missions. Regardless of the letter designation used, the spacecraft are numbered sequentially after they are launched. Thus, Nimbus C became Nimbus II after its launch in May 1966. The Nimbus II mission is similar to the Nimbus I. However, there are some significant technological improvements in the basic spacecraft, notably, in the attitude control system, where some of the gains in the individual control loops have been adjusted for optimized performance, in the power supply area, and in the data system. There are two additional experiments on Nimbus II beyond the three flown on Nimbus I. The first experiment is simply a combination of two earlier components: namely, the HRIR and the APT. The Nimbus II spacecraft transmits the high-resolution infrared data through the APT link during the nighttime as well as TV pictures

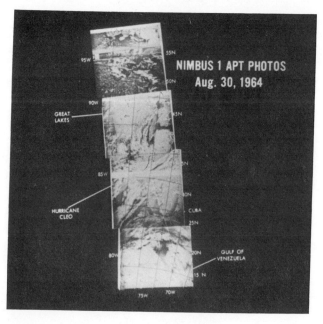

Figure 13   Series of single pass pictures showing cloud formations from the Great Lakes to Venezuela.

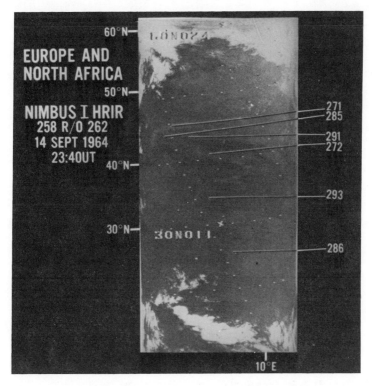

Figure 14   Facsimile display of HRIR covering the Mediterranean area.

during daylight hours, so that the local user in the United States will receive TV cloud-cover pictures at noon and infrared cloud pictures at midnight. This will allow a significant improvement in local weather forecasting.

We have also added an additional experiment called the MRIR (Medium Resolution Infrared Radiometer), which is a five-channel radiometer with a resolution of 30 mi. This experiment will generate data to evaluate the global heat balance, the earth's heat budget. This experiment has flown on at least one of the TIROS satellites, but this is the first time that global data can be collected since TIROS was inclined at a much lower angle, and its data collection was only intermittent.

As of the latter part of July 1966, the Nimbus II spacecraft had completed more than 60 days in orbit and was functioning very well. As an example of the volume of pictures and general meteorological data

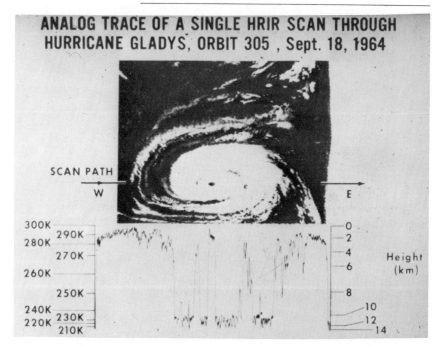

Figure 15   Analog trace through Hurricane Gladys.

produced by the spacecraft, the following sensory data were collected from Nimbus II in its first 60 days in orbit:

AVCS	63,978 pictures
HRIR	630 hr
MRIR	1146 hr
APT	666 hr
Direct readout	
HRIR	430 hr

The third of the Nimbus meteorological spacecrafts, the Nimbus B, represents a major change in configuration. The accent on Nimbus I and Nimbus II (in fact the emphasis on all meteorological satellites) has been on cloud pictures of one kind or another, primarily in the visible region, from which a great deal of useful meteorological information has been derived. However, there are hopes that more meaningful meteorological information can be developed and more direct quantitative measurements can be taken. It is to this end, that the seven new experiments on Nimbus

B are directed. These include an infrared spectrometer and an infrared interferometer spectrometer, which is quite similar to the spectrometer; the IRLS interrogation and ranging location sytsem, which is a data collection system, similar to the French EOLE experiment as described by Dr Morel in Part 7. The technological portions of these systems will be tested on Nimbus B. In addition, an ultraviolet experiment is on Nimbus B. This is a sun-looking experiment and attempts to correlate the intensity of two selected UV wavelengths with rocket data taken from the ozone region above the earth to try to test the theory that there is a cor-relation between the sun's radiation in the UV and the ozone equilibrium. There is an improved high-resolution infrared radiometer and a single camera intended to replace both the AVCS, that is, the stored mode camera, and the APT. This camera uses a very old TV sensor in a new way. The image disector is the first TV sensor ever developed. Its lack of any storage capability presents no problem on Nimbus, since in stored mode, the system stores the data on tape and in the real-time APT mode, there is no storage. The image disector is an extremely simple sensor, with no filament and very few parts. It holds a great deal of promise for ulti-mate operational use because the limited lifetimes of Vidicon sensors are the critical elements in most camera systems. Finally, a technological experiment on Nimbus B is an RTG (Radioisotope Thermoelectric Generator) power supply using plutonium 238. This system is called SNAP-19 and consists of two SNAP-19 generators, each one nominally rated to deliver 30 W of power.

The Nimbus D mission which was mentioned above, is in the formative stages as of 1967. The focus of Nimbus D will be to study the atmosphere further and to be ready to accept experiments being proposed now.

In summary, Nimbus is a meteorological observatory which serves to flight test new sensors for operational use. Nimbus has been the father of almost all the meteorological sensors now flying. The Nimbus AVCS camera is on TIROS and TOS and will fly on the ATS (Applications Technology Satellite) program and the APT camera has flown fairly widely. The Nimbus mission in addition to providing the capability for a test laboratory for new and important sensors, has developed an ad-vanced attitude control system, a unique power supply and improved data systems. Finally, of course, the Nimbus program represents a challenging job to those of us who work on it with the knowledge that our results can contribute to better weather forecasting and, hence, to the welfare and happiness of everybody on earth.

VERNER SUOMI

Verner Suomi was born in Eveleth, Minnesota, and received a B.A. degree in 1938 from the Winona Teacher's College and a Ph.D. degree from the University of Chicago in Meteorology in 1953. Since 1948, Dr. Suomi has been on the faculty of the University of Wisconsin and has been a full professor since 1958. He is married and has three children.

Abstract—*In December 1966, the spin-scan camera designed by Dr. Suomi and Professor Robert Parent was put into a synchronous orbit over the Pacific. The camera takes cloud-cover pictures every 20 min of a large portion of the earth. From these stationary snapshots Dr. Suomi has made motion pictures which show the changes in the cloud patterns and the development of storm systems. Such a visual record is very useful in working toward and understanding of the global circulation patterns.*

Dr. Suomi ends Part 5 with a discussion of the spin scan cloud camera which is being flown on the ATS (Applications Technology Spacecraft) series of satellites.

The Spin Cloud Camera

AT the time these lectures in the Systems Course at Stanford University were given, ATS-1 had not yet been launched. My lecture merely described the main elements of the system, since some of the details were not yet worked out. The camera performance in orbit on ATS-1 exceeded what we had expected. Many, many highly useful photographs of the earth's disc over the Pacific are now available.* In order to up-date this article I have used parts of reports written by Roger Thompsen of the Santa Barbara Research Center, who built the camera, and also the reports of the ATS project under Robert Darcey, the project manager. I wish to acknowledge the many contributions of a large and dedicated team and especially my colleague Professor Robert Parent, who made the original signal-to-noise calculations.

ATS-1, a 775-lb spin-stabilized spacecraft, was launched successfully from the Eastern Test Range on 6 December 1966. After injection into an equatorial synchronous orbit (22,300 mi above the earth) the spacecraft was oriented so the camera could view the earth. The first picture was obtained on 9 December 1966. The camera data are transmitted on the wide-band communications channel from the satellite to the ATS ground stations at Rosman, N.C., and Mojave, Calif. At the ground stations the video signals and sun-pulse information from the spacecraft telemetry system are combined and high-quality photographs prepared. In addition, the data are processed and recorded on digital tapes at Rosman and on analogue tapes at Mojave.

A requirement of the spin-scan cloud-cover experiment is to continuously monitor the weather motions over a large fraction of the earth's surface. Even though near-earth weather satellites have provided an impressive array of visual and infrared observations of the earth's weather on a nearly operational basis, the synchronous satellite affords another

* ATS Spin-Scan Catalog available from NASA.

opportunity to gain a better understanding of the *global weather circulation,* the key to *better weather prediction.*

The view from a near-earth satellite is so fleeting that it is not possible to obtain any real measure of the *weather motions.* For example, in TIROS series of satellites, the life history of a model storm had to be derived from a number of different storms, at different times, at different places, and in different stages of development. A synchronous satellite allows one to measure the cloud motions rather than infer them because the earth's disc is under continuous surveillance. Figure 1, which is a view of the Western Hemisphere taken on 21 January 1968, shows the global

Figure 1   A view of the earth from synchronous satellite altitudes showing global cloud patterns.

nature of the cloud patterns. The Americas are on the right side of the picture.

In the tropics the weather motions have a shorter time scale than the motions at higher latitudes. The tropics, between $\pm 30°$ latitude, covers half the earth's area which is 80 percent ocean. Here the surface observations are very sparse and polar-orbiting satellites have the greatest gaps in their data.

The tropical region is the "boiler" of the giant atmospheric heat engine. Convective activity plays an all-important role in the heat-transfer process, yet its short scale prevents its being observed adequately by near-earth satellites.

A high-resolution synchronous meteorological satellite can provide needed spatial distribution information of cloud systems over a large fraction of the tropics and higher latitudes as well. Even more important, it can provide key information on cloud-system changes over time scales smaller than the diurnal interval. Finally, since the camera system provides dimensionally accurate plots, it can provide information on cloud displacement as well, which in turn provides information on the large-scale air motions which transport the heat out of the tropics to higher latitudes. A synchronous meteorological satellite is a key tool in this study. In all that follows, the capability of extracting the cloud motion from a series of pictures has been of paramount importance in making the choice of the system to be used. From a synchronous orbit it is possible to take "time exposures" of the weather motions on the scale of the earth's disc. This permits one to use exceedingly simple electronic and optical systems to generate high-resolution photographs. A 2-mi resolution at the subsatellite point has been demonstrated. Several alternative camera systems suited to the ATS spin-stabilized spacecraft were considered. They are:

(1) An image-dissector camera tube using various combinations of lenses or reflective optics.*

(2) A moving "pin hole" in the image plane of a 7-in. focal length Kodak Ektar F 2.5 lens. The light passing through the "pin hole" is measured with a photomultiplier tube after first having passed through a minus blue filter to improve the contrast.

(3) A "rocking telescope" system which employs a 10-in. focal length Cassegrain telescope whose primary mirror is 5 in. in diameter. A stationary "pin hole" aperture has a minus blue filter and a photomultiplier detector similar to system 2.

---

* An image dissector has been flown on ATS-III launched in November 1967.

In each of the camera systems listed above, the east–west scan is generated by the rotation of the spacecraft. The north–south scan must be generated in the camera system. One can do this magnetically by deflecting the electron image along the spin axis using an image dissector tube (System 1), the aperture can be moved along the spin axis (System 2), or the whole camera can be tilted on an axis normal to the spin axis (System 3).

IMAGE DISTORTION

Detrimental image distortion can arise from two sources: deficiencies in the camera itself and nutation of the satellite spin axis.

(A) Camera Distortion

As already mentioned, one of the most attractive features of the synchronous satellite for meteorological purposes is that the weather moves, not the satellite. The east–west scan provided by the spinning satellite is highly linear and extremely stable, so that east–west geometry will be highly precise more or less automatically. If a highly precise north–south scan can also be provided, it will be possible to determine cloud motions by a comparator technique similar to that used in astronomy to obtain the proper motions of the stars. Lens or mirror distortion is not as important as sweep distortion, since lens distortion is fixed, but sweep distortion may be dependent on temperature, voltages, and so on. This feature is so important meteorologically that it counted heavily in the choice of the telescope camera system. In system 3 on-axis optics is used everywhere in the photo so no optical distortion exists.

(B) Picture Distortion Due to Nutation of the Spin Axis

Spin axis misalignment will cause only a trivial amount of image distortion. On the other hand, spin axis nutation will cause serious picture degradation, but not total data loss. However, experience with ATS-1 shows that nutation is no problem whatever. Even following a spacecraft orientation maneuver the nutation damps out in a few minutes. After an hour or so any residual nutation is below the noise level of the sensitive nutation detection accelerometer.

The spin-scan "camera" is not a camera at all. It consists of a high-resolution telescopic photometer with a photomultiplier light detector. The telescope is coupled to a precision latitude step mechanism. This latitude step motion combined with the spinning motion of the ATS spacecraft provides complete coverage of the earth from 52.5° north latitude to 52.5° south latitude and from the west limb to the east limb. This area is scanned by 2000 horizontal (west to east) lines. The optical resolution is 2.2 nmi when the telescope is pointed at nadir from a synchronous equatorial orbit, 22,752 nmi from the earth's center. A general view of the geometry is shown in Figure 2. The major components of the spin-scan camera system in addition to the telescope and its precision step mechanism which provides the north–south scan, are a sun sensor from which synchronizing signals are generated, the spacecraft communication system and the ground station synchronizer, display system and associated magnetic tape recorders. The main camera specifications are listed in Table 1.

The camera optical telescope is a two-element reflective system com-

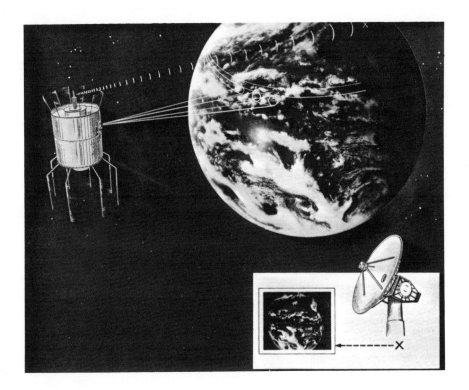

Figure 2 General view of the operation of the spin-scan cloud camera on board the ATS.

TABLE 1
CAMERA SYSTEM PARAMETERS

1. Optical System
Type	two-element reflective
Primary mirror	5-in. parabolic
Secondary mirror	2-in. diameter flat
Surface	Aluminum coated with Mg fluoride
Mirror substrate material	Fused quartz
Instantaneous field of view	0.1 ± 0.02 mrad diam. $\frac{1}{2}$ power point
Aperture	0.001 ± 0.0001 in. diam.
Focal Length	10 in.
Filter	GG-7
Spectral bandpass	4750° to 6300 Å

2. Scan
Line Scan	Spacecraft rotation (100 rpm)
Lines per frame	2000 ± 50
Scan limits	$\pm 7.5°$ ($\pm 52.5°$ lat.)
Scan drive	Step scan, one step per rev. of S/C. Provided by sealed mech. drive
Frame time—normal	20 min
Frame time—back-to-back	20 min
Retrace time—normal	2 min
Retrace time—back-to-back	20 min
Scan timing source—normal	PACE and 10 × PACE

3. Electronics
   Video Amplifier
Gain stability 0° to 50°C	± 5 percent
Linearity	± 2 percent
Dynamic range	$\leq 1000/1$ (video ampl.)
Electronic upper cutoff frequency	200 kHz (-3 dB)
Electronic lower cutoff frequency	0.1 Hz (-3 dB)
Outputs Output 1	Video + sun pulse
Output 2	Video 1 less 10 dB + sun pulse

4. Command and Signal Inputs
   From Spacecraft
Step camera—scan	1 per revolution from PACE 158° after camera has scanned past N-S earth centerline
Step camera—retrace	10 per revolution from PACE
Sun-pulse	1 per revolution when in sun (approx. 2 msec duration)

TABLE 1 (*cont'd*)

4. Command and Signal Inputs (*continued*)	
From Ground through S/C	
Normal scan mode	N-S Scan, 20 min; S-N retrace 2 min
Back-to-back scan mode	N-S Scan, 20 min; S-N scan, 20 min
North scan limit	Either mode ± 5 lines
South scan limit	Either mode ± 5 lines
5. Power	
Power input to system (max)	−24 V dc, 1 A
Temp. monitor	−10 V dc
6. Size	
Outline	10 × 11 × 7 in.
7. Weight	
Maximum	≤ 20 lb less cabling
8. Spacecraft Voltage Controlled Oscillator	
Input voltage	1.0 V P–P
Maximum deviation	± 7.5 MHz
Center frequency	65.39 MHz or 66.33 MHz
Output linearity	10 percent deviation from a best fit line
Input impedance	75 Ω

posed of a 5-in. diameter fused quartz parabolic primary and a flat 1.8-in. diameter secondary. The equivalent focal length is 10 in. The optical surfaces are evaporated aluminum overcoated with magnesium fluoride. The effective area of the entrance aperture is 104 cm^2. A field-defining aperture is located at the focal plane and provides a circular instantaneous field of view of 0.1 mrad diameter. This field aperture consists of a 0.001-in. diameter hole in a titanium-gold evaporated film on a quartz substrate. Quartz was chosen for its excellent resistance to damaging radiation (UV, IR, and so forth), titanium for adherence properties, and gold for high reflectivity. Intermittent exposure to focused solar radiation requires these properties.

Energy collected by the optical system and focused on the field aperture passes through this aperture and spreads over the 1-in. diameter photo-cathode of the photomultiplier tube through a diverging lens. The lens also provides the required blue cutoff filtering because it is made from a special Corning filter glass.

This f/2 optical system required the use of an invar structure and fused quartz optical mirror substrates to provide the necessary dimensional stability with temperature. A compensating type of structure could have been used with an associated reduction in weight; however, expected non-

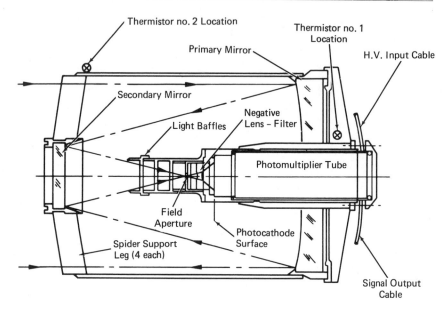

Figure 3 Spin-scan cloud camera optical configuration.

uniform thermal gradients and a tight program schedule ruled out this approach.

In operation the entire camera is constantly exposed to a force of approximately six times gravity owing to spacecraft rotation. This constant load required that the optics be hard mounted if defocus was to be avoided. Mounting surfaces were carefully lapped to prevent optical surface distortion by the retaining forces. The telescope layout is shown in Figure 3.

The major mechanical components of the spin-scan cloud camera are the main housing, the telescope, and the precision step mechanism.

The housing is a lightweight dip-brazed aluminum structure composed of two halves to provide ease of assembly. Final machining is done with the halves bolted together for accurate location of the mounting surfaces. The two electronic areas are completely accessible from the outside and have lightweight protective covers.

The telescope consists of a tubular structure made from rolled sheet stock, a spider assembly with secondary mirror holder, and a photomultiplier tube housing. All parts are made from Invar with the exception of the aluminum light baffled aperture housing. The telescope housing parts are dip brazed prior to final machining of the mirror mounting surfaces.

Two Bendix flexural pivots support the telescope in the main housing and allow for the limited 7.5° telescope motion. The flexural pivots have the advantage that they are nonlubricated bearings with no radial play. Any radial play in the telescope bearings would introduce error in the step scan linearity and step position repeatability. Step position tests on both the prototype and flight model cameras affirm the choice of the flexural pivots. The complete camera assembly is shown in Figure 4.

Even though the rate at which pictures taken by the ATS spin scan camera is slow indeed when compared with ordinary TV practice, the requirements for stability in the ground station synchronizing circuits are

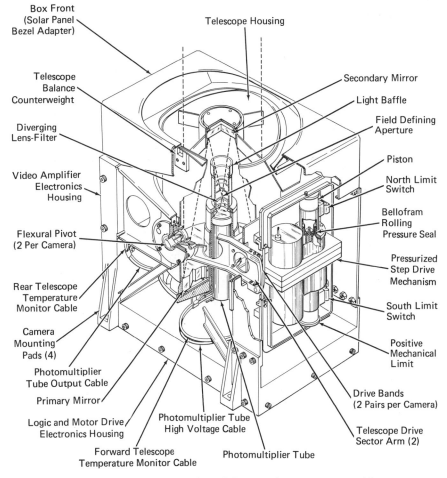

Figure 4   Cutaway drawing of the complete camera assembly.

very severe. If the spin rate of the spacecraft were constant, an oscillator which generates a pulse for each turn of the spacecraft would need to have a stability of 1 part in 10^8. The spin rate of the spacecraft is obtained by a sun sensor of the spacecraft. Thus the spin rate is measured against the *sun*. What is desired is a spin rate is measured against the *earth*. The spacecraft measured against the earth makes one revolution per day fewer or greater than if measured against the sun, depending on the direction of rotation. The sun pulse drift in the 20 min required for a full picture amounts to about one-third of the horizon-to-horizon East–West scan distance. Even though the sun pulses are transmitted to earth over the wide-band telemetry link, they may contain enough noise to interfere with proper synchronization. For example, the sun pulse, whose duration is about 2 msec, will last about 1/15th of the duration of the video burst which occurs when the camera scans the earth along the equator. We can expect the sun pulse to have a gaussian shape with an uncertainty in amplitude and duration of about 10 percent, or 200 μsec. The timing accuracy on the ground needs to be held to 10 μsec on any one scan line. However, the 10^8 stability arises from the requirement that this error must not be exceeded over 2000 individual scan lines which take a total time of 20 min. The 10^8 stability requirement applies only to the long-term frequency stability for the whole picture and not for each line. The stability requirement for one line need be only one part in 3000.

Because the telescope motion shifts its mass moment about the spacecraft spin axis slightly, the spin period of the satellite is not constant but changes about 110 μsec out of the 600-msec spin period. This seemingly small change would accumulate to a large error if a constant frequency oscillator were used in the ground system. Instead a phase lock system, operating on the sun pulse, is used to provide the synchronizing signals.

Since the smoothed sun pulse generated by the phase locked loop occurs at the time the sun sensor sweeps past the center of the sun, additional timing signals must be generated in the ground synchronizing system in order to display the received earth video. In Figure 5 the angle B which must be determined is the angle through which the satellite rotates between passage of the sun sensor over the sun and the camera sweep over a given earth reference line. This angle B is the angle between the projections of the satellite sun line and satellite earth line into the satellite spin plane. Since this angle varies with time, a means of providing an initial B and $\dot{B} = dB/dt$ must be provided. $\dot{B}$ depends on the sun and satellite positioning, satellite spin axis orientation, time of year and any satellite motion with respect to the ground. The computation of B and $\dot{B}$ and the effect of these parameters on picture resolution is not discussed in this

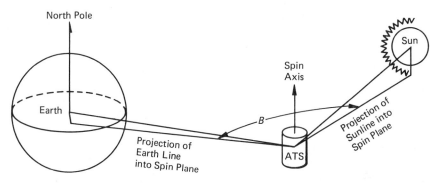

Figure 5   Satellite rotation earth–sun reference angle *B*.

paper; however, the method of using these parameters with the timing, subsystem will be discussed.

On the ground the timing subsystem receives the smoothed pulses from the phase-locked loop and generates most of the timing signals employed by the various recorders. The functional block diagram is shown in Figure 6. The initial information for the sun-earth angle is computed

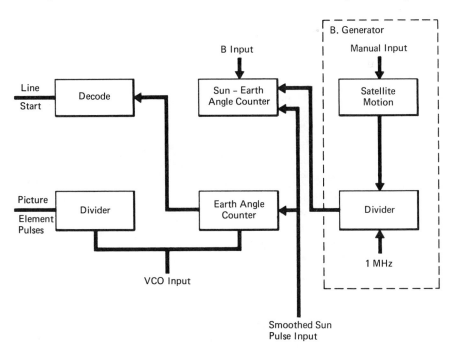

Figure 6   Camera timing subsystems.

from the satellite ephemeris data and is manually set into the sun-earth angle counter at the beginning of each picture frame. The number in the sun-earth angle counter thus is directly proportional to the current angle B between the sun and the earth. The smoothed sun pulse, generated in the phase-locked loop, is used to transfer the contents of the sun-earth angle counter into the earth angle counter. The earth angle counter counts the clock pulses from the phase-locked oscillator so that it cycles through zero at a fixed angle with respect to the earth. Additional timing control signals are generated by decoding selected counts of the earth angle counter. They include line start, line end, and picture element pulses. The contents of the sun-earth angle counter are incremented at a rate controlled by the signal input sun-earth angle rate $\dot{B}$, keeping it continuously proportional to the angle B.

The recording system consists of an analog, digital, and an electronic photofax recorder. The photofax recorder operates from real time data and produces on Polaroid 550N film a positive print and a negative. The print is basically for monitoring purposes and the negative is of high resolution and quality for subsequent enlargements. The photofax recorder receives the analog video from the video processor and timing signals from the synchronizer. The video information is then displayed on a high resolution 5 in. cathode ray tube and projected on the film. The mechanism for the tube deflection is basically a digital one. The recorder has both X and Y deflection. The X deflection corresponds to the azimuth sweep and the vertical deflection corresponds to some particular elevation of the camera. The deflection waveform is generated by two digital counters—one for hoizontal and another for the vertical whose numerical output, after digital to analog conversion, drives the deflection amplifiers. The net effect is like that of locating a given cell on cartesion-coordinate graph paper by giving its numerical coordinates. Once a cell has been located, a gate turns on the cathode ray tube (CRT) electron beam. The light given off when the electrons hit the CRT phospher is divided by a half-silvered mirror into two components, of which one is directed to the film and the other to a photomultiplier. The photomultiplier integrates the light output and continuously compares its reading with the brightness dictated by the incoming video signal for that cell. When the two values coincide, a gate turns off the electronic beam and the system is ready to move on to a new cell. This process is continued until the complete picture is generated.

The horizontal deflector register has 12 bits, or 4096 states, and the vertical deflector register has 11 bits with 2048 states. Thus, 4096 uniformly spaced spots can be placed in a horizontal line, and in the vertical

Figure 7   Synchronous satellite cloud cover picture taken on 18 November 1967 showing the Atlantic Basin, South America, and the western part of Africa.

direction a total of 2048 lines. Intensity information can be presented from analog video or 8-bit digital data for each spot. The exposure capabilities of the recorder are accurate to roughly 2 percent of the video level and 16 shades of gray may easily be displayed.

The advantage of the CRT imaging system just described is that it can operate in real time directly from the spacecraft signals. The 5 percent duty cycle is improved by first digitizing the video signals in a fast A-D converter, storing this data one line at a time in a core memory, then

Figure 8   Photo of Baja, California, using photo facsimile recorder operating from analog tape signals. Guadalupe Island is visible in the hole in the clouds.

sorting it out at a slower rate on the CRT display. The CRT limits the resolution and the linearity.

A somewhat similar system has been developed by Hughes Aircraft Corp., but instead of a CRT a photo facsimile machine is used. The drum rotation is made to be identical with the spacecraft spin rate by a phase-locked oscillator system. The duty cycle is improved by core storage as in the Westinghouse EIS system. Here the resolution is limited by the spot size of the light source and the mechanical precision of the scanning drum and lead screw. This system operates in real time also.

Finally, the video signals are recorded on magnetic tape, each video burst being stretched to occupy about 85 percent of the time rather than the 5 percent of the time it actually takes. This is done with a recurrent scanner analog tape recorder. This recorder has a drum phase locked

172 7 2215

Figure 9   High resolution photofax copy of the full earth showing detailed cloud patterns.

to run at four times the spacecraft spin rate. The record head is gated to record only when the earth video signal is being received.

These duty cycle improved signals are then displayed on a facsimile machine modified to provide the required number of scan lines.

Figures 7 to 9 show photos of sections of the earth using these systems and also some enlargements to show the cameras true performance.

PART 6

the atmospheric models

CECIL LEITH

Cecil Leith received a B.S. degree in mathematics in 1943 from the University of California at Berkeley. From 1946 to 1968 he worked for the Lawrence Radiation Laboratory, until 1952 in Berkeley on nuclear scattering experiments and after 1952 at the Livermore Laboratory on the construction of numerical models of various physical systems. Since 1968 he has been at The National Center of Atmospheric Research in Boulder, Colorado. He received the Ph.D. degree from the University of California, Berkeley, in 1957 in mathematics, specifically, linear operator theory. Dr. Leith became interested in numerical models of the atmosphere in 1958 and started work on the model discussed here in 1960.

Abstract—*Dr. Leith discusses the history and development of numerical modeling of the atmosphere. He discusses in general the problems he has encountered in developing his six-layer model.*

A Six-Level Model
of the Atmosphere

T HE problem of predicting future states of the atmosphere from a given state is a very old one. In some ways it is the oldest of scientific problems, since people have been worrying about being able to predict the weather for several thousand years. It was recognized as long as 60 years ago that, in principle, this was a problem which could be solved in much the same way as future orbits of planets could be predicted in that all the fundamental laws which influence the way in which states of the atmosphere evolved were well known. The hydrodynamists of the 19th century had derived all of what seemed to be the appropriate physical laws, and so the question naturally arose whether this problem was not just a difficult calculation which, given enough time and enough patience, could be solved directly.

In fact, in 1911 the first effort was made by the British mathematical physicist, L. F. Richardson, to carry out such a calculation. He did not have available to him at that time the computing facilities that we now have, and he worked on the calculation by hand; but the procedure that he used, or at least the plan he laid out, was much the same as we now try to progam on high-speed computers. He divided the atmosphere over England and Europe into boxes, a mesh. He started with available values of temperature, water vapor, and winds, and he used the four equations of conservation of mass, momentum, water-vapor substance, and energy to set up a calculational cycle which would predict, three hours later, what the new state of the atmosphere would be.

This single three-hour time step that he was trying to compute, in fact, took him some four or five years of calculation, which he carried on through World War I during his spare time as an ambulance driver at the front. Evidently this was not a technique that was going to be immediately applicable to weather prediction. But he did want to test whether, in principle, one could carry out such a calculation based on

205

some finite difference approximation to the equations of hydrodynamics and get a reasonable answer. He found an answer, finally, on a change in surface pressure at some point in England, but because this figure turned out to be about 60 times greater than had ever been observed, it could be looked upon as a dismal failure. After all, he was trying to see whether he could get a reasonable answer from such a calculational process and he got an unreasonable one.

This naturally had the effect of discouraging people from pursuing this further for some 30 years until shortly after the Second World War when computing facilities became available which were quite a bit more powerful than working with paper and pencil. In the late 1940's a group at the Institute for Advanced Study at Princeton, having access to the computer, the Maniac, which was built under the guidance of John von Neumann, tackled this problem anew. Then, of course, they realized that if something was going to go wrong, it would go wrong soon enough to allow them to try something else or to fix it instead of waiting five years to find out that they had not succeeded.

In the interim, the reasons for Richardson's difficulty had become clearer as more thought was applied to this problem. The first difficulty was that he had chosen a finite difference appoximation to the differential equations describing the conservation laws that for reasons of numerical stability demanded a time step, not of three hours but of five or ten minutes. However, this discrepancy by itself would not have caused him any trouble until he got to the 20th or 30th time step in the calculation. Reaching this point would have taken so long at the rate he was computing that it is not of any real consequence here.

The other problem, which was more serious and which is still with us today, was that the initial data he had used were based on the observations available to him and inevitably contained a certain amount of error. In particular, they had a fairly large amplitude of what is now referred to as "meteorological noise" (errors in the measurement of the initial data). This is a mode of behavior of the atmosphere which does not normally occur, but which is supportable and describable by the equations used and which corresponds to the presence of gravity waves of fairly large amplitudes—the sort of waves occasionally induced in the real atmosphere by explosions of volcanos. This noise phenomenon is not generally present in the atmosphere, and for a number of reasons, not completely understood. The equations used in Richardson's models permitted this gravity wave motion to exist, so that within three hours all the errors in the initial observations had moved very rapidly some place else, as gravity waves do, leading to the very violent fluctuation in the distribution of surface pressure that he computed.

Recognizing these difficulties, von Neumann's group, in the late 1940's, began to compute the evolution of the atmosphere's behavior with a model of the atmosphere so simplified that this particular type of gravity wave motion could not exist. In this model the atmosphere became a two-dimensional, incompressible fluid flowing like an ocean over the surface of the earth so that the motion of the atmosphere could be described by one single dependent variable, namely, a stream function, from which the velocity field could be deduced. In this simple model there was no consideration of what the vertical distributions of temperature or water vapor were; this was what we refer to as a one-level model of a very restricted sort. It was initially used to describe the evolution of the flow of the atmosphere over the United States and later was extended to cover the Northern Hemisphere. It did surprisingly well. Within a very few years it was possible to predict by these numerical techniques the evolution of the flow of the atmosphere over the Northern Hemisphere with an accuracy (or inaccuracy) equal to that of the traditional weatherman with his experience and intuitive feel for weather.

In the mid 1950's there was a group set up by the Navy, the Air Force, and the Weather Bureau to try to convert this research work into an operational technique for weather forecasting. About five years ago the Weather Bureau started using numerical forecasting techniques for predicting the behavior of the atmosphere over the Northern Hemisphere. Since this technique was doing as well or a bit better and was more objective than human forecasters, it began to be widely used. It had long been observed that a meteorologist's skill at forecasting increased markedly during his time on the shift, whereas the numerical model, on the average, did equally well or equally poorly every day. The mathematical approach was, in addition, using the computer, the differential equations of motion, and the basic laws of hydrodynamics—a more modern way of doing things.

An unfortunate historical accident occurred at this time. The use of numerical models was becoming more popular at just the time when perhaps the greatest boon to human forecasters was becoming available, namely, satellite photographs of cloud cover. These pictures are extremely difficult to convert into the kind of numbers appropriate for the initiation of a numerical prediction, but they permit a human forecaster to make a better judgment of the state of the atmosphere at a particular time and to make use of his great pattern recognition ability in improving his forecasting (see Mr. Elliot's article in Part 2 on analog methods of forecasting). A person who has worked with computers for any length of time develops a very good feeling for the types of things people can do better than the computer; pattern recognition seems to be one of these.

Within the last five years the amount of effort spent on far more elaborate models of the atmosphere has increased. Such models eventually will be used for forecasting purposes, although they are at present still in their infancy because they are so much more elaborate than those being used for the current 18-to-36-hour forecasts. These models are set up for the purpose of making very long-term calculations and integrations for many days, weeks, or months, rather than the short-term forecasting calculations for a three-day period. If one wishes to carry out a calculation for many weeks or months, then one has to take into account energy sources and sinks, frictional interaction with the surface of the earth, and other influences that have only a slow effect on the behavior of the atmosphere. These are the effects that can be, and have been, ignored in the short-range, two- or three-day forecasting models.

The reason for interest in these long-term calculations is that one would like to set up models that would simulate atmospheric behavior over long periods of time to understand the forces that influence the climate. For a long-term calculation the problem becomes one of deciding how the boundary conditions being imposed on the model influence the statistical behavior of the atmosphere. This may then lead to answers for long-range forecasting but not answers of the specific sort that short-range forecasting tries to provide, for example, that it is going to rain two days from now. Long-range forecasting will probably never answer the question of whether it is going to rain on a certain day three months from today, but it may well be able to answer such questions as whether next winter will be unusually wet. Very long range forecasting will make statistical rather than specific weather predictions.

The model we have been working on in Livermore was set up about seven years ago to make optimum use of the LARC computer which was delivered in late 1960. One of the major problems in this field is to make the very best use of available computers, since this whole process of setting up numerical models of the atmosphere is straining the speed and size of all computers. I should say first that the LARC computer is of a speed somewhat greater than an IBM 7094 and somewhat less than the IBM STRETCH computer. It has now been superseded by faster computers, not only the STRETCH, but the CDC 6600, which is perhaps a factor of 5 to 10 faster than the LARC.

Professor Mintz, from the University of California in Los Angeles, has been working with a model very similar to ours which describes the vertical structure of the atmospheres with two levels. The model we have been using in Livermore uses six. The vertical coordinate in this and most other models is pressure rather than height. The atmosphere is divided vertically into six pressure levels or layers so that various dependent variables are kept track of at the 100-, 200-, 400-, 600-, 850-, and 1000-mb levels of pressure.

At each of these pressure levels one records two quantities: the temperature and the amount of water vapor (the water-vapor mixing ratio). At the centers of each of the pressure levels the two horizontal components of velocity (the horizontal wind components) are recorded. This array, then, over any mesh point corresponds to a total of 24 different numbers for the six levels used to describe the vertical wind and thermal and water-vapor structure of the atmosphere. There is, in addition, a 25th number in this particular coordinate system. Since pressure is the independent variable, there is a dependent variable which describes the position of the lower pressure surface which now becomes a free surface boundary at the earth's surface. This bottom surface pressure, one of the oldest of meteorological variables, is the variable recorded on a barometer and is a measure of how much of the mass of the atmosphere has accumulated over a given point on the surface of the earth. The model then uses 25 such numbers over any particular horizontal mesh point.

The horizontal mesh used covers the globe with intervals of 5° in latitude and longitude, which corresponds to a mesh size of 300 mi or 555 km at the equator, although, or course, it is a function of latitude. It is so variable that at 60°N the longitudinal dimension has been reduced by a factor of 2, and to avoid singularities it is necessary to widen the angular measure of the mesh as one goes toward the poles.

The one vertical and two horizontal dimensions are three of the independent variables, and there is, of course, a fourth variable—time. There is a time axis in these calculations divided into 10-min intervals. In other words, we carry out a calculation which advances the variables over a time step of 10 min.

Now I mentioned that what Richardson had done, and what we do here, is to use a set of equations which is a finite difference approximation to the equations that describe the conservation laws of energy to advance the temperature to the next time interval, of water-vapor substance to advance the water-vapor mixing ratio μ (mu), and of momentum to advance the wind-velocity components.

The conservation law for mass is used in the following special way. An approximation was necessary before one could use pressure as a vertical coordinate; namely, that the atmosphere in the model is always in hydrostatic equilibrium. The differential relation between pressure and height becomes $dp = -pg\ dz$ (p is the mass per unit volume relative to the x, y, z coordinate system). This means that in this particular model, relative 'to this coordinate system, the atmosphere is an incompressible fluid. A "volume" element in this p, x, y coordinate system is

$$dp\ dx\ dy = -pg\ dx\ dy\ dz = -gdm$$

which is simply $-gdm$ or g times the mass of the element. In other words, a "volume" element always contains a certain amount of mass and in this coordinate system, conservation of mass guarantees incompressibility. Thus the model atmosphere behaves very much as if it were an ocean which differs from the water ocean only because its bottom rather than its top surface is free. Variations in surface pressure describe the free bottom boundary.

The conservation laws for energy, water vapor, and momentum are all directly related to the existence of sources and sinks in the atmosphere, for neither energy, nor water vapor, nor momentum are conserved identically. There are energy sources and sinks which drive the atmosphere and make it move in the way that it does. There is evidently more energy coming into the atmosphere in the equatorial region, and there is a loss or a deficit of energy in the polar regions. This source-sink situation converts the atmosphere into something of a heat engine which tries to generate from the sources and sinks the mechanical motion of the winds. The most important energy source is the release of latent heat of vaporization which occurs upon condensation of water vapor. Another important energy source is the absorption of incoming solar radiation by the water vapor found in various layers in the model. There is also a certain amount of heating from the small-scale convective transport of heat off the surface up into the atmosphere. Thus these three—heat of condensation, solar radiation, and convective heating—are the major sources of energy in the atmosphere, and all are calculated explicitly in the model.

Of course if there were only such sources, then the atmosphere would continue to increase its temperature indefinitely. There must be a sink of energy to maintain the observed balance in the atmosphere. The primary energy sink is the outgoing terrestrial radiation computed in the model explicitly. Terrestrial radiation is the effective blackbody radiation at the earth's temperature which is in the infrared part of the spectrum, as contrasted with solar radiation (sunlight), which is blackbody radiation at the sun's temperature. Terrestrial radiation is determined by the temperature as well as by the amount of water vapor found at different levels. The water-vapor content is again important because terrestrial radiation is absorbed and remitted rather readily by water-vapor substance, and the atmosphere, at least in certain wavelength intervals, is relatively opaque to terrestrial radiation.

The water-vapor conservation equation also has sources and sinks. The source is evaporation from the ocean surfaces which brings water vapor into the atmosphere (this is computed explicitly in the model); the sink is the condensation process, which in turn was an energy source. In this model as in the atmosphere there is evaporation from the surface of the ocean; the water vapor may then be carried to a new location, and finally as there are changes in state such that part of the atmosphere becomes saturated, condensation occurs, latent heat energy is given up, the water-vapor content is decreased, and liquid water is removed from the atmosphere. In short, it rains. This cycle was also quite important in the energy cycle of the atmosphere, and in this sense the atmosphere may be considered to be a steam engine.

The momentum equations also have sources and sinks associated with them. These are the frictional interactions with the surface which can fluctuate and which are taken into account in the model.

Unfortunately, technical difficulties in connection with finding the finite difference approximations to the conservation equations exist. From the discussion thus far it may seem that one need set up only an initial condition, start integrating these equations in 10-min time steps on into the indefinite future, and develop some average behavior of the model, which hopefully will be similar to the real atmosphere. In fact, it took some time to find numerical finite difference approximations to these equations which would continue in a regular way over an indefinite period of time, and a few numerical problems still remain. I distinguish these from meteorological problems having to do with certain aspects of the original equations about which we are still uncertain.

There is a rather basic point in connection with these numerical models which in some ways is the essential difficulty. It is clear from the coarseness of the mesh that the model can describe explicitly only scales of motion comparable to or larger than the mesh size. It is fairly clear that, given only one value of the wind every 300 mi, the model cannot possibly describe the fine structure associated with frontal activity, which is, perhaps, 50 mi in scale, or the events associated with convective processes such as single clouds or thunderstorms, which are perhaps 10 mi across. Evidently these phenomena cannot be described in such a coarse model as this. Yet the atmosphere has all these scales of motion present in it, and energy is being transferred from one of these scales to another at all times in this turbulent flow. One faces the twofold dilemma of determining the influence of these smaller scales and motions on the larger scales of motion and, inversely, of accounting for the energy cascaded from the larger scales of the model to the smaller scales. The equations describing the evolution of the atmosphere are nonlinear, and all the problems of turbulence theory arise.

We deal with this serious complexity numerically by introducing diffusion terms which have a smoothing effect on the motions of the atmosphere. Once one recognizes the scope of this problem there is ample justification for adding these largely artificial diffusion terms to the equations. Theoretically one may estimate an eddy diffusion coefficient to describe the effects of the smaller scales of motion on redistributing quantities, but then the problem of choosing a value for the diffusion coefficient remains. Finally, the value for this diffusion coefficient comes from experiment, and the justification for it comes from whatever turbulence theory arguments can be made. We have thus taken a very pragmatic view by simply introducing enough smoothing or damping to let the calculation run indefinitely. This is a serious problem for the horizontal small-scale processes (dimensions less than 300 mi).

The atmosphere, however, is by no means so extensive in the vertical direction; the scale happens to be only 10 km. Vertical small-scale motions exist, which are also quite important in the vertical transport of water vapor and heat. One of the present problems in these models is to improve the vertical convective processes which are transporting heat and water vapor from one layer to another. This problem has arisen only now that we are pretending to describe the atmosphere with many layers. When we had only one or two layers, we did not worry so much about the vertical transport of quantities between them. But now that we have six, if we are going to use them properly, we should actually try to compute the transport of such quantities vertically as well as horizontally. This last problem has not yet been completely solved and is one on which much effort at present is being expended.

The lastest work has led to an inversion of the usual procedure in using models. One would initially like to start with known basic laws of physics, and using a computer, put all these together into a large complicated system to show the evolution of complex states of the atmosphere. This has been done quite well for the conservation laws. However, at present we simply do not understand the physical laws which govern many of the turbulent convective processes in the atmosphere. We must thus reverse the procedure to move from the observed behavior of the atmosphere to an understanding of the statistical laws relating the small-scale convective processes to the average values of temperature, water-vapor diffusion, and their gradients.

These calculations with this degree of refinement and with all the effects which I have described take 1 minute on the LARC computer for each 10-min time step. That means that this global model runs 10 times faster than the real atmosphere, which is just barely fast enough for forecasting purposes; it would be better to have a ratio of 100 between computing time and real atmosphere time. Such a factor will be soon available, with careful use of computers such as the CDC 6600 which now exist and with not so careful use of computers such as the CDC 7600, which will be available within another two years.

The problem of removing from the model the noise component or measurement error which plagued Richardson is still a problem, especially in using general equations for the entire globe. I have not faced this problem in this model, since for initial conditions I have used a state of the atmosphere which is completely artificial but which is free of noise. I have set up a model state which does not correspond to any that ever existed in the real world, in which the north–south component of the wind is zero (the flow is only from east to west) and in which everything is a function only of latitude, not of longitude. This is a zonal flow situation. For such a flow it is rather easy to compute what the temperature, water vapor, and wind fields should be in order for them to be internally consistent and free of noise components. From such a completely artificial state of the atmosphere, computation proceeds, and after about a month of atmosphere time, the typical behavior of the model begins to show up. After a month, for two or three more months it has an irregular but somewhat cyclic behavior, which is characteristic of the model and more or less characteristic of the real atmosphere. After about three or four more months, because of an improper balance of some of the energy sources, which are now becoming important, the temperature distribution in the atmosphere becomes somewhat erratic and the behavior ceases to be realistic.

Many years ago it was pointed out that a certain paradox existed in the search for an accurate numerical model of the atmosphere, for, should it be found, its behavior would be just as complicated and just as little understood as that of the real atmosphere. In what sense could it then be said that the equations of atmosphereic motion had been solved? The answer could be that an algorithm had been found that permits in any particular case the prediction of resulting behavior. The algorithm is, however, unusually time consuming and expensive to carry out, and the results are too complex to be readily comprehended.

The model state vector has some 50,000 components, and the listing of these as a function of time is overwhelmingly noninformative. Much effort is being expended on this information problem; the natural first step is to represent the fields of meteorological variables as contour maps similar to the surface-pressure maps published in the daily newspaper. Experience has shown that one can thus display the information contained in about 1000 components as a single map (in agreement with an early estimate credited to Confucius). By superposition in different colors two or three such maps may be displayed together, and finally by use of successive maps as motion picture frames a time-lapse motion picture can be produced. Even all this technique involves the selection of only a part of the total information contained in the model. It is enough, however, to show that the model simulates many of the observed features of the real atmosphere.

The results of the calculation are seen to be, then, animated color cartoons of the behavior of a more or less realistic model of the earth's atmosphere. If this were all, it would seem an amusing but expensive toy. The more important benefit is the ability to carry out experiments with the model that would be impossible or irresponsible with the real atmosphere. These can determine the relative importance of various external influences on the behavior of the atmosphere. The model also serves to exhibit by its limitations those areas of significant ignorance (such as moist convection) for which the rewards of further research may prove the greatest.

APPENDIX

REPRESENTATION OF STATE FUNCTIONS
AND NONLINEAR (TURBULENCE) EFFECTS*

The first question that arises in the planning of a numerical model of fluid flow concerns the most suitable numerical representation for the functions describing the hydrodynamic and thermodynamic state of the flow. Any choice represents of necessity a compromise, for any finite representation must be an approximation to the infinite amount of information characterizing the true state.

If the equations describing the evolution of the flow were a set of linear differential equations, then linear theory suggests that a most natural choice would be a representation of the state functions by expansion on a basis of characteristic functions of the spatial differential operators involved. (To describe a flow confined to a spherical surface, for example, it would be natural to use surface spherical harmonics as a basis.) The original partial differential equations would be transformed thus into a set of ordinary time differential equations describing the evolution of the expansion coefficients, and in fact the state at any time could be computed as a sum of contributions from independently evolving components. This computation lacks being a complete solution to the problem in that we are restricted to a finite and incomplete basis. We know (and perhaps need to know) only the solution in a finite dimensional manifold, but the manifold is invariant and we can safely ignore the rest. We might call this process of decomposing the function space into a known and unknown manifold *linear truncation*.

Unfortunately the equations for the flow include important nonlinear terms, and the methods of linear analysis lose much of their effectiveness. If one proceeds as for the linear problem to make an expansion in functions characteristic for the linear spatial derivatives, then the nonlinear terms are trans-

---

* This Appendix and the last three paragraphs of Dr. Leith's article are drawn from the report by C. E. Leith, *Numerical Hydrodynamics of the Atmosphere* Livermore, Calif: Lawrence Radiation Laboratory, March 1966. All references cited in this Appendix will be found in the original report.

215

formed into nonlinear sum relations linking the evolution of the coefficients. No longer is it possible to find an invariant finite dimensional known manifold without artificially cutting off the links to the unknown manifold. Thus further cutting off might be called nonlinear truncation. Such "wave-number space" representations have been found useful in many investigations of the hydrodynamics of the atmosphere.

Perhaps the most common representation of the state functions is by their evaluation at a finite number of mesh points more or less uniformly distributed throughout the space-time domain of the original differential equations. The partial differential operators are then approximated by finite difference expressions, reducing thereby the integration of the evolution equations to a numerical process. It is this approach which is followed in the model described above.

This configuration space representation has clearly also defined a known manifold in the function space; the unknown manifold contains functions which vanish at the mesh points. Unfortunately even for the linear terms this known manifold is no longer invariant and there remain the further troubles of nonlinear interaction between manifolds. It would seem then that the configuration space approach has little to recommend it. Its principal virtue is that the nonlinear terms are differential and thus local in configuration space, while being nonlocal in wave number space. Nonlocal terms involve for their evaluation many more arithmetic operations than local terms, the ratio of arithmetic requirements being proportional to the dimension of the known manifold. This typically can be 10^5.

It has been pointed out that the configuration space mesh should be more or less uniformly distributed. No problem arises here if the spatial domain is a rectangular box, for then each dimension can be subdivided uniformly into a mesh of equal intervals in the three space coordinates. Unfortunately the domain of the flow of the earth's atmosphere is more nearly a spherical shell. The choice of distribution of mesh points over the surface of a sphere involves further compromise. A natural choice of horizontal coordinates is latitude and longitude; the difference equations have a simple form for equal mesh intervals in these coordinates. This leads, however, to singularly narrow mesh zones in the polar region. Not only does this waste computing power, but can for common and efficient choices of difference equations impose a too severe restriction on the choice of mesh time interval. The mesh which has been used for the model described here divides the region between 60° north and 60° south latitude into a horizontal mesh of 5° spacing in latitude and longitude. Between latitudes 60° and 75°, however, there is a coarsening of the longitudinal mesh to an interval of 10°. Between 75° and 80° the longitudinal interval is 20°, and finally from 80° to 90° the interval is 40°. Such a choice introduces discontinuities in the mesh at the transition latitudes of 60°, 75°, and 80° which complicate the difference equations and introduce new sources of error. The pole is surrounded by nine triangular zones and requires special difference equations.

Another approach to this problem used by Smagorinsky has been to carry out a coordinate transformation corresponding to a polar stereographic projection for, say, the northern hemisphere. Subdividing the projected map into a square mesh leads to reasonably uniformly shaped mesh elements with a factor-of-two variation in linear dimension between the pole and the equator. For global coverage it is necessary to use also a southern hemisphere projection and to carry information from one mesh to another at the equator by an interpolation process. This interpolation process is probably the weakest link in the procedure, but this technique has been used successfully.

Another variant, from the work of Phillips, has involved the similar patching together of polar stereographic projections for the polar regions to a Mercator projection for the equatorial zones, the connection being made at middle latitudes.

The most recent and potentially the most satisfactory solution to the mapping problem is one by Kurahara, in which the discontinuous transitions of the first scheme are smoothed out so that each latitude circle has its own choice of longitudinal interval. The resulting mesh leads of course to more complicated finite difference equations and thus to more arithmetic operations.

The specification of the mesh intervals in the vertical direction is more straightforward. Although the earth's atmosphere extends tenuously indefinitely outward, the part of it of greatest concern is where most of its mass is to be found. It is convenient and a good approximation for global models to assume that the atmosphere is in hydrostatic equilibrium. This means that the pressure at any particular point is due only to the weight per unit area of the air above that point and permits pressure to be used as a vertical (masslike) coordinate. In this model the vertical mesh division is into six pressure layers. Such a coordinate scheme has removed the problem of an ill-defined upper boundary for the atmosphere, but it has replaced it with the problem of computing the position (that is, pressure) of the lower boundary, the earth's surface, which now becomes a free boundary changing in space and time.

It has already been pointed out that nonlinear terms in the differential equations of hydrodynamic flow lead to an interaction between the known and unknown manifold in a wave-number space representation, and that this difficulty is not removed by using a configuration space representation.

In the case of a numerical model of the atmosphere we may view this difficulty as being due to the turbulent nature of the flow. It is clearly impossible to define and compute all motions on all scales. The detailed motion of a dust devil whirling down an alley in Calcutta or of the air currents swirling about a mountain peak in the Sierra Nevada of North America must remain undescribed in any global model. In fact with a horizontal mesh interval of the order of 500 km, we can only hope to explicitly describe scales of motion of horizontal dimensions of the order of 1000 km and larger. The influence of the smaller scales of motion on the behavior of the larger explicitly computed scales must be treated statistically. Turbulence theory in its present state gives no clear answer on how this should be done—only a few suggestions which are being followed.

In a three dimensional isotropic and homogeneous turbulent flow the effect of nonlinear terms is to permit large-scale motions to interact in such a way that there tends to be a transfer of energy to smaller scales. Any finite difference model of such flow should also lead to such an energy cascade, but now a natural limit is imposed by the mesh interval on how far this process can be properly computed. It has been observed that in finite difference models energy transfer to scales outside the describable range does not simply vanish from the system but rather, through an "aliasing" process, reappears erroneously as an increase in energy at the low-wave-number (large-scale) alias of the proper wave number. The low wave numbers are hopefully those of greatest significance in the calculation, and the first problem is to try to prevent their being contaminated by this aliasing error. To the extent that the nonlinear processes are removing energy from describable wave numbers, the effect may be likened to a dissipation, and it is tempting to introduce an "artificial" viscosity term in such a way that it will remove that amount of energy which in fact is being removed by nonlinear transfer processes. Such an approach has been taken by Smagorinsky and Lilly. (The original report contains further discussion of the form of this viscosity coefficient.)

An aspect of the influence of unknown small scales on known large scales which is specifically encountered in atmospheric models is that of convection. A familiar sight under certain atmospheric conditions is the towering cumulus convection cells of thunderstorms. These transport upward vast quantities of heat and moisture. Yet being only a few kilometers in size, they evidently cannot be explicitly described by the model. The largest single energy source to the whole atmosphere is the release of latent heat in convection cells in the tropics. Unless this source can be computed with reasonable accuracy, the model must remain inaccurate. In particular in models with a resolution of the vertical dimension into many layers the vertical transport by convection can be far more important than by explicitly described motions in determining the vertical temperature and moisture distribution. The solution to this problem consists in knowing the statistical laws for average convective transport given as functions of the large-scale average moisture, temperature, and wind distributions. Such statistical laws are not yet well known from either theoretical prediction or observation.

The theoretical problem of thermal convection has been worked on continuously since Rayleigh's solution of the problem of the onset of Bernard cell convection. Recent work on convection in a single fluid such as dry air has extended the theory into the more pertinent domain of fully developed, large-amplitude, nonlinear convection. Convection in the atmosphere is further complicated by the possible phase change from water vapor to liquid with associated latent heat release. A linear theory of such moist convection has been worked on recently by Kuo.

While waiting for the solution to this problem, it is possible to formulate some likely convective laws which include a few undetermined parameters. Such tentative laws can be introduced into the numerical models and the parameters adjusted until the model agrees with the real atmosphere in its behavior. There is, of course, the danger in this procedure that one can obtain the correct behavior for the wrong reasons. It is clear that the complicated behavior of the atmosphere is no longer being deduced from a set of simple fundamental physical laws but rather the fundamental laws are being induced from the observed behavior.

YALE MINTZ

Yale Mintz received a B.A. degree from Dartmouth in 1937, an M.A. degree from Columbia in 1942, and the Ph.D. degree from U.C.L.A. in 1949. He is now Professor and Chairman of the Department of Meteorology at the University of California, Los Angeles, and also Professor of Planetary Science at Tel Aviv University, Israel. Using numerical methods, Dr. Mintz has done extensive research on the general circulation and climates of planetary atmospheres.

Abstract—*Part 6, The Atmospheric Models, constitutes one of the most important topics covered in this book, since the global satellite observing systems will supply the initial data from which the weather forecasts using numerical models will be made. In his paper, Dr. Mintz describes the four basic requirements for numerical weather prediction. He discusses the physical laws, mathematical and computational techniques, speed of computing, and the global weather observations necessary to solve this initial value problem.*

The Four Basic
Requirements for Numerical
Weather Prediction

To predict the weather by numerical methods, we must (1) know the physical laws which govern the atmosphere on the scale of the weather systems, (2) have adequate finite difference approximations for the differential equations in which the laws are expressed, (3) specify the initial state of the dependent variables, and (4) carry out the numerical computations faster than the weather develops.

The following is a discussion of these four requirements.

1. THE PHYSICAL LAWS

For the scales of atmospheric motion in which the horizontal dimension is much larger than the vertical dimension, we can assume hydrostatic balance, so that the pressure is equal to the weight of the overlying air. With this approximation, the behavior of the atmosphere is governed by the following laws: the law of hydrostatic balance, expressed by the hydrostatic equation,

$$\frac{\partial z}{\partial p} + \frac{1}{g\rho} = 0 \tag{1}$$

the law of conservation of horizontal momentum, expressed by the horizontal component of the equation of motion,

$$\frac{\partial \mathsf{W}}{\partial t} = -\mathsf{W} \cdot \nabla_p \mathsf{W} - \omega \frac{\partial \mathsf{W}}{\partial p} - 2\Omega_z \cdot \mathsf{W} - g\nabla_p z + \mathbf{F} \tag{2}$$

the law of conservation of energy, expressed by the thermodynamic energy

221

equation,

$$\frac{\partial T}{\partial t} = -\mathsf{W} \cdot \nabla_p T - \omega \left(\frac{\partial T}{\partial p} - \frac{RT}{C_p p} \right) + \frac{h}{C_p} \tag{3}$$

the law of conservation of mass, expressed by the continuity equation,

$$\nabla_p \cdot \mathsf{W} + \frac{\partial \omega}{\partial p} = 0 \tag{4}$$

and the gas law, expressed by the equation of state,

$$\rho - \frac{p}{RT} = 0 \tag{5}$$

where W is horizontal velocity, T is temperature, ρ is density, and p is pressure (which is used as the vertical coordinate). $\omega \equiv (dp/dt)$ is individual rate of pressure change, and is approximately proportional to the downward vertical velocity. Ω_z is the vertical component of the earth's rotation vector, R is the gas constant for air, C_p is the specific heat of air at constant pressure, and g is gravity. ∇_p is the horizontal gradient operator in the surface of constant pressure, z is isobaric height, and t is time. F is the horizontal component of the frictional force per unit mass, and h is the diabatic rate of heating per unit mass.

This set of equations for the large-scale motions of the atmosphere is called either the "hydrostatic system of equations" or, more commonly, the "primitive equations."

The solution of these equations requires specification of the upper- and lower-boundary conditions. The upper boundary of the atmosphere can be taken where the density and the pressure are zero, so that

$$\omega \equiv \frac{dp}{dt} = 0 \qquad (\text{at } p = 0)$$

At the lower boundary of the atmosphere, the air flows parallel to the ground. From this condition and the equation of mass continuity, we obtain the surface pressure tendency equation

$$\frac{\partial p_s}{\partial t} = -\nabla \cdot \int_0^{p_s} \mathsf{W}\, dp \tag{6}$$

where p_s is the pressure at the earth's surface and ∇ is the horizontal gradient operator.

The three prediction equations, Equations (2), (3), and (6), determine the evolution of the variables with time.

The equation of horizontal motion, Equation (2), tells us that the local time rate of change of the horizontal velocity of the air depends on the horizontal and vertical advections of horizontal velocity, on the horizontal acceleration by the Coriolis force, on the horizontal acceleration by the pressure gradient force, and on the horizontal acceleration by the frictional force.

The thermodynamic energy equation, Equation (3), tells us that the local time rate of change of the temperature of the air depends on the horizontal and vertical advections of temperature, on the adiabatic temperature change due to the individual pressure change, and on the diabatic heating of the air.

The surface pressure tendency equation, Equation (6), tells us that the local time rate of change of the surface pressure depends on the divergence of the vertical integral of the horizontal mass transport.

If the frictional force F and the heating h are known, this is a closed system of six equations in six unknowns, which is to be solved as an initial value problem. This can be done by prescribing the distributions of the horizontal velocity W, temperature T, and surface pressure p_s, at an initial time t_0. Then, integrating Equation (4) with respect to pressure, from $p = 0$, will define ω. Substituting from Equation (5) into Equation (1), and integrating Equation (1) with respect to pressure, from $p = p_s$, where the height is known, will define the isobaric height z. The right-hand sides of the three prediction equations, Equations (2), (3), and (6), can then be calculated, giving the local time rates of change of W, T, and p_s at the initial time. From these time rates of change, and an appropriate integration procedure, we obtain the change over an increment of time Δt; and hence W, T, and p_s, for the time $(t_0 + \Delta t)$. Repeating the procedure N times gives us the prediction of the fields of W, T, and p_s, for time $(t_0 + N\Delta t)$.*

Other variables can be added to this system. We can add the components of the composition field $q_i = (q_v, q_w, q_r, q_o, q_n, \ldots)$, where q_v, q_w, and q_r are water substance in its vapor phase, water droplet phase, and ice crystal phase; q_o is atmospheric ozone; q_n are the sublimation or condensation nuclei; and

$$\frac{\partial q_i}{\partial t} = -W \cdot \nabla q_i - \omega \frac{\partial q_i}{\partial p} + \dot{q}_i \tag{7}$$

* Instead of prescribing the initial distributions of W, T, and p_s, we could prescribe the initial velocity W and density ρ. If the density is given as a function of height from the top of the atmosphere downward, then integration, with respect to height, of the hydrostatic equation, in the form $\partial p / \partial z + g/\rho = 0$, will define the pressure. Introducing the density ρ and this hydrostatically computed pressure p into Equation (5) will then define the temperature T.

is the equation of continuity for each component of the atmospheric composition. This equation tells us that the local time rate of change of each composition component depends on the horizontal and vertical advection of the component and on its sources and sinks q_i. The composition field enters the system through its influence on the atmospheric heating $\dot{h}$.

As discussed further in the next section, we can solve only the finite difference approximations of these differential equations. In doing so, the continuous atmosphere is replaced by a finite number of discrete variables. The physical quantities are carried at discrete points in space and the space derivatives are replaced by space differences.

The discretization produces two kinds of errors. One kind is the errors that come from the omission of atmospheric processes which are on a scale too small to be resolved by the chosen finite difference grid. These are the "physical errors" of discretization. The other kind, which we will discuss in the next section, is the "mathematical errors" of discretization.

The only way of reducing the physical errors of discretization is to parameterize the statistical behavior of the subgrid scale physical processes and use these parametic functions in the governing equations for the grid scale system.

Thus, in the equation of motion for the grid scale horizontal velocity W, the subgrid scale processes would be parameterized in the frictional force F. In the thermodynamic energy equation for the grid scale temperature T and in the several continuity equations for the grid scale composition components q_i, the subgrid scale processes would be parameterized in the diabatic heating $\dot{h}$ and in the sources and sinks for the composition $\dot{q}_i$.

A simple example would be the parameterization of the vertical transfer of horizontal momentum by subgrid scale motions by making the transfer proportional to the vertical gradient of the grid scale wind and a parameter that varies with the vertical gradient of the grid scale temperature. In this formulation, the temperature lapse rate would control the vertical momentum flux (the vertical "eddy stress") in a manner analogous to the way in which the grid of a thermionic valve controls the flow of electrons from anode to cathode. But this simple formulation could produce errors; for example, in regions where the cumulus motions of the atmosphere are so organized that they might transport horizontal momentum upward *against* the gradient of the large scale wind.

In particular, it is important to have a good parameterization of cumulus convection, especially in the tropics. The latent heat release by tropical cumulus convection is a principal heat source for the general circulation. Therefore, we must have a good parameterization of the statistical properties of cumuli as a function of the grid scale variables. In addition to the latent heat release by cumulus convection, we need to parameterize its vertical transports of sensible heat, horizontal momentum, and composition components.

Other subgrid scale physical processes which need to be parameterized are the turbulent transfers of momentum, heat, and moisture through the atmospheric boundary layer. This must be done, not only to calculate the direct atmosphere-earth interactions, but also to calculate the boundary-layer wind which largely controls the moisture supply for the areas of cumulus convection.

In addition, visible and infrared radiation, in clear and cloudy air, affect the temperature field. The radiative transfers are particularly sensitive to clouds, and hence the subgrid scale cloud fields must be parameterized.

Finally, the atmosphere and the underlying earth are a coupled system. In particular, the atmosphere is coupled to the oceans. The dominant influence of the oceans upon the atmosphere is through the ocean surface temperature, as this influences the heat and water vapor flux at the air-sea interface, which in turn affect the temperature and circulation of the atmosphere. On the other hand, the dominant influence of the atmosphere upon the oceans is through the circulation. The surface wind stress produces ocean currents, upwelling, and mixing of the ocean boundary layer, all of which affect the ocean surface temperature, and in this way the circuit is completed.

Over the continents, the coupling between the atmosphere and the underlying earth depends mainly on the available ground moisture and on the snow cover. These are the principal time-dependent properties of the ground, and through the surface evaporation and reflection of solar radiation they affect the composition field and the heating of the atmosphere.

2. FINITE DIFFERENCE APPROXIMATIONS

It was pointed out above that when the continuous atmospheric fields are replaced by values of the variables at discrete points, errors will result from the consequent omission of physical processes too small to be resolved by the chosen grid. We indicated how those "physical errors" of the discretization might be reduced, to a smaller or greater extent, by parameterization of the statistical properties of the subgrid scale processes in terms of the grid scale variables.

The second class of errors may be called the "mathematical errors" of the discretization. These come from solving the difference equations instead of the differential equations. The difference equations must be chosen so that these errors are as small as possible.

The primitive equations, given in Section 1, govern two types of motions: one is the hydrostatic high frequency gravity-inertial waves, the other is the low-frequency motions seen on weather charts. Both types of motions are quasi-horizontal, but the low-frequency motions are also quasi-nondivergent.

As a result of the "geostrophic adjustment mechanism," most of the energy of the atmosphere is in the low frequency motions, and there is relatively little energy in the inertia-gravity waves. As a minimum requirement, the finite difference scheme must properly simulate the geostrophic adjustment mechanism which confines the energy to the low-frequency waves. For this particular purpose, there is no need for the scheme to be highly accurate, though it must be computationally stable. But a high order of accuracy is necessary for properly simulating the behavior of the low-frequency motions, which determine the weather.

The low-frequency atmospheric motions are governed by nonlinear dynamics, in which energy is transferred from one part to another within the spectrum of the large-scale motions. Therefore, it is important that the finite difference scheme be nonlinearly as well as linearly stable, especially if we wish to carry out long-term integrations. If the finite difference scheme approximates the constraint on quadratic quantities, such as energy, the scheme will be stable. One of the authors (Arakawa, 1966) has shown how this can be achieved through the design of the advection terms, which are the principal nonlinear terms in the equations.

The finite difference scheme should also be relatively free of computational modes, false dispersion of waves, and false boundary and internal reflections. Not all these problems have been completely solved, but progress is being made.

It is also desirable to have the distance between grid points quasi-constant over the globe. Such a grid would be quasi-homogeneous in accuracy; and it will have no strong gradients of grid size, which can cause false internal reflections of energy. Because the shortest grid distance, which determines the time step for stability, would be close to the average grid distance, this would also be an economical grid to use. An example of such a scheme, based on the icosahedral-hexagonal grid, has been given by Sadourny, Arakawa, and Mintz (1968).

3. SPECIFICATION OF THE INITIAL STATE

With a knowledge of the physical laws, an adequate finite difference scheme, and sufficient computing capability, we can simulate the atmosphere in "numerical experiments." When integrated over long periods of time and in the hemispherical or global domains, these are called "numerical general circulation experiments." In a numerical general circulation experiment, one can begin with an arbitrary initial state: for example, a state of rest, isothermalcy, uniform composition and uniform surface pressure; an initial state of random fields of the dependent variables; or the like. Given the arbitrary initial state, the time integration is then carried forward long enough for the atmosphere to respond fully to the nonlinear dynamical processes, and to the heating and friction functions, to reach a long-term, or statistically steady state (which is not usually a stationary state), in which the motions and other characteristics of the atmosphere have become relatively independent of the initial state.

Such numerical general circulation experiments are very useful. They show us the extent to which the numerical model can simulate the climate; that is to say how well the model can reproduce the statistical properties of the atmosphere (the time averages of the variables, the fluctuations, cross correlations and spectrum of the variables, and so on, down to the characteristic structures and behavior of the individual atmospheric disturbances).

But, for an actual weather prediction, we must begin with the real initial state, based on observations. An error in this initial state will usually grow and cause the prediction to deteriorate with time. Even if there were no physical or mathematical errors, the growth of the initial state error limits the predictability (which we can define in terms of the time it takes for the difference between the predicted state and the true state to become as large as the characteristic difference between any two randomly chosen states, for the same time of the year.)

The growth of the initial state error may not be very rapid. Some global predictability experiments, by Charney (1965), have indicated that initial root-mean-square errors of the temperature field have a doubling time of about five days, so that given a $1°C$ root-mean-square error, the limit of deterministic predictability is about 15 to 20 days. However, beyond this time interval, there may still be some useful nonclimatological statistical information in the forecast.

When integrating the primitive equations, the time-dependent variables that must be initially defined are W, T, p_s, and q_i. However, this does not mean that the observation of each variable must be equally accurate; nor does it even mean that all the variables must be observed. As was pointed out, in Section 2, the primitive equations govern two types of motions: high-frequency gravity-inertia waves, which have little energy, and low-frequency motions, which have most of the energy and affect the weather.

Except in areas of inertial instability, which are usually limited to small regions, the geostrophic adjustment mechanism produces an approximate balance between the mass field (given by p_s T) and the wind field $\mathbb{W}$. But it is not practical to use prognostic equations which explicitly maintain this balance condition (the so-called filtered equations), if we wish to obtain the same high physical accuracy that we get with the primitive equations. However, we can use this balance condition for determining the initial state. With it, we can compensate for the lack of observations, or inaccurate observations of the initial wind and mass fields. For example, in the middle latitudes we can compute the wind from the mass field (assuming that we have measured the mass field accurately) probably as well as we can measure the winds. This, however, is not true in the tropics.

On the other hand, if we can measure the entire wind field accurately, then (assuming we know the vertical distribution of the horizontally averaged pressure) we can compute the temperature field and the surface pressure field over the entire globe.

Both of the above procedures will depend on knowing the field of the observed quantity fairly accurately, and also knowing the physical processes of friction and heating accurately. For practical reasons, therefore, it should be better to use a mixed system, in which both the mass wind and the fields are measured, with an emphasis on the wind field in the tropics.

We have learned from the predictability experiments, that for predictions up to 15 to 20 days it is necessary to know the initial temperature field to 1°C. The approximate balance between the wind and the mass fields gives us an equivalent accuracy requirement of about ± 2 m sec^{-1} for the initial wind, and about ± 3 mb for the initial surface pressure.

It is difficult to determine the comparable accuracy requirement for the components of the composition field q_i. Of the three phases of water substance, the mass is larger in the vapor phase than in the clouds of liquid droplets and ice crystals, and therefore the residence time of the water substance is larger in the vapor than in the liquid and ice phases. Consequently, although it may be easier to observe the state of the liquid water and ice clouds, it is more important to measure the initial field of water vapor. A rough estimate of the required accuracy of the water vapor measurement, consistent with the above requirements for the other state variables, is ± 1 mb for the vapor pressure.

We are unable to estimate the accuracy requirement for the other components of the composition field.

The horizontal and vertical resolution of the observations must also be considered. If we wish to predict the motion and weather systems on the scale of a few thousand kilometers, we must also carry shorter waves, perhaps down to a few hundred kilometers wavelength, in order to allow for possible energy cascades from the motions we wish to predict to smaller-scale motions. This means that we need a horizontal resolution of the order of 100 km, and a corresponding vertical resolution of about 100 mb in the troposphere. If we add some additional levels in the surface boundary layer and in the stratosphere, we must have about 15 levels in all, or a total of 10^6 grid points in the domain.

This resolution, of about 100 km between grid points in the horizontal, with 15 levels in the vertical, will be the *computational* resolution, and it will determine the computer requirements discussed in the next section. However, the *observational* resolution does not necessarily have to be as high as the computational resolution.

The observational grid only needs to be fine enough to get good resolution of the initial state of the motions we want to predict, those on the scale of a few thousand kilometers. The consequent errors in the initial state of the smaller-scale motions will be tolerable, as the residence time of the energy is normally smaller in those motions. Our guess, therefore, is that we can tolerate observing the initial state with a half to a quarter of the resolution used in the computations; that is to say with about 200-to-400-km horizontal resolution and, correspondingly, from 8 to 4 levels in the vertical. This will make the number of observational points of the order of 10 to 100 times fewer than the number of computational points (or 10^5 to 10^4 observational points compared with the 10^6 computational points). The initial state variables at the intervening computational points would then be parameterized from the statistical relationship between the variables on the observational and the computational scales. The point of view being expressed here is: if the larger-scale fields are correctly represented in the initial state, but the smaller (100-km scale) features are missing or are otherwise incorrectly represented, the smaller features will soon form in more or less the correct way by internal adjustment of the meteorological fields, and this will not cause any large error of the prediction.

Because we are solving an initial value problem, it would be convenient to have simultaneous (synoptic) observations. But this is not absolutely necessary for the low-frequency motions. New observations can be introduced continuously, as corrections to the predicted fields, at the computation times corresponding to the individual observations. It should be sufficient to introduce these corrections with the local frequency of once a day. Each new global prediction for the period from t_0 to $(t_0 + N \Delta t)$, would then start as a calculation from t_0 minus one day, where t_0 is the time of the latest observation.

4. COMPUTER REQUIREMENTS

The computing speed for carrying out the numerical computations faster than the real weather develops depends on several factors: the number of arithmetic and logical operations per grid point per time step, the number of grid points in the domain, the length of the time step, and the desired gain of computing time over real time.

The number of arithmetic and logical operations, per grid point per time step, depends on the finite difference scheme and on the degree of complexity of the parameterization of the subgrid scale physical processes. For most schemes, with relatively simple forms of parameterization of the subgrid scale processes, a characteristic number of operations is about 2×10^3 arithmetic and logical operations per grid point per time step.

As we have already indicated, a desirable computation resolution is about 100 km horizontal spacing between the gridpoints, with about 15 levels in the vertical. The extent of the domain depends on the duration of the forecast. For short-range predictions, up to about three days, a hemispherical domain is sufficient. But for longer-range forecasts the domain should be global. The global domain, 15 levels, and a quasi-constant horizontal mesh size of 100 km, requires 10^6 grid points.

The time step is determined by the requirement for computational stability. With the primitive equations, the maximum Δt must be of the order of the grid size divided by the speed of external gravity waves. For a grid size of 100 km this is 5 min.

For operational weather predictions, a practical ratio of computing time to real time is about 1 to 20; that is to say, about 1 hour computing for a 1-day forecast, or half a day for a 10-day forecast.

When we combine the 2×10^3 operations per grid point per time step, 10^6 grid points, 5 min time steps, and a 1 to 20 ratio of computing time to real time, we have 10^8 operations per second as the required computing speed. This about 100 times faster than the speed of the best existing digital computers. But computers have been designed which will have even greater speeds than 10^8 operations per second. One of these, the SOLOMON computer at the University of Illinois, is under construction.

With respect to computer errors, there is data roundoff due to the finite word length. But if care is taken in programming, a 32-bit word length is sufficient to make the roundoff error negligible, compared with the other errors.

There are also intermittent machine failures. But as these are almost always catastrophic in their effects, they are easily identified. They require only that some intermediate results be stored for restarting economy.

PART 7

the second stream

STANLEY RUTTENBERG

Stanley Ruttenberg was born in St. Paul, Minnesota, and grew up in Philadelphia and Pittsburgh. He studied physics at Johns Hopkins University and the Massachusetts Institute of Technology, where he received the B.S. degree in 1946. He became a teaching assistant and graduate student at the University of California in Los Angeles, receiving a M.S. degree in physics in 1952. From 1949 until 1955 he worked at the Institute of Geophysics in upper-atmospheric research and from 1955 until 1964 he was a member of the staff of the National Academy of Sciences, Washington, D.C., working primarily on aspects of the International Geophysical Year (IGY) and the International Years of the Quiet Sun (IQSY). His duties included editing and preparation of IGY reports as well as participation in the production of the IGY film series "Planet Earth"; a series of highly successful color films for secondary schools. Since 1964 Mr. Ruttenberg has been the Assistant to the Director of the National Center for Atmospheric Research at Boulder, Colorado.

Abstract—Dr. Hallgren's concept of the "second stream" includes the programs and weather systems which will bring the World Weather Watch to maturity. Mr. Ruttenberg begins this chapter with a description of the U.S. GHOST balloon system. He suggests that the major obstacle to an understanding of the general circulation of the atmosphere is our current lack of global atmospheric data. And he proposes the GHOST system of free-floating weather balloons as the first stage of the solution to this data shortage.

The GHOST Free-Floating Balloon System

THERE are at least three prerequisites for an effective long-range weather forecast:

(1) A global numerical model of the atmosphere.
(2) Measurements of temperature, pressure, wind velocities, water-vapor content, and so forth from several levels of the atmosphere from the entire globe to feed into the numerical model as initial state data.
(3) An effective communication system to transmit predictions to all potential users.

The purpose of this article is to discuss the need for initial-state global weather information.

One of the main reasons why we cannot make accurate long-range weather predictions now is that we have insufficient data on which to base the predictions. At any one time we can observe at most 20 percent of the atmosphere, and this simply cannot provide enough data for the numerical models to make accurate predictions of more than three to four days.

There is good observational coverage in North America, Europe, and parts of Asia, from the standard meteorological networks; there are a few spotted observations in oceanic areas and a few weather ships operated for the United States by the Coast Guard; there is information from merchant ships which travel regular routes; and there is information from commercial aircraft. We have information from satellites, but at the moment it is difficult to interpret cloud pictures and the infrared radiation patterns to derive features of the general circulation. Our greatest gaps are in unpopulated parts of the world and the oceanic areas, and it seems clearly impractical if not technically impossible to expand the existing methods of obtaining meteorological information to cover these areas.

That is why many groups have given attention to a scheme for obtaining meteorological information over the oceanic areas that will supplement, not replace, the standard meteorologic network.

Obviously, the weather bureaus of the United States and other countries require better data in order to make better forecasts. They have operational needs for data up to the level of the jet aircraft, say 300 to 200 mb, but there is not as much need for data above 200 mb. In the United States, however, we regularly obtain data to 10 mb and occasionally to 5; a few European countries also regularly obtain data to 10 mb. Such data are not very useful so far for short-term forecasting in the lower atmosphere, but it has become clear in the last few years that information at the high levels is important for long-term forecasts in the lower atmosphere. In oher words, the longer ahead you want to forecast, the higher up in the atmosphere and the better distribution you need, because small effects become important after several days. The models, such as that described earlier by Dr. Leith, when given only input data from the lower atmosphere, seem to do quite well for several days and then begin to diverge rapidly from the real atmosphere. We do not know yet all the reasons why they diverge and what data and techniques we need in order to maintain their parallelism with the real atmosphere, but we do know that something is wrong, part of which is the lack of data over the entire world.

Good global data over a reasonable depth of the atmosphere, for a reasonable length of time, say six months, would be invaluable for scientific research. With a good set of data, meteorologists could study many phenomena. It would not matter that the data were 10 or 20 years old; or for that matter, it would not matter whether all the data were real, if at least scientists had some real initial conditions to work with. Our current models, according to some of the theoreticians and the operations people, are limited by the lack of real data. From two fronts, then, there is a very strong need sometime in the next five years to expand our coverage of the world by a factor of 5 to 10, from the present 10 to 20 percent, to full coverage of the world. Many global weather systems have been proposed, and many more no doubt will be suggested before a worldwide system becomes operational. The rest of this article will concentrate on one promising system—the GHOST (Global Horizontal Sounding Technique) balloon system, on which NCAR has been working in cooperation with the U.S. Weather Bureau for the past five years.

Figure 1 (which is identical with Figure 5 in Dr. Hallgren's article) is a map of the world on which are plotted the upper-air radiosonde stations; many, but not most, of these stations also have radio tracking

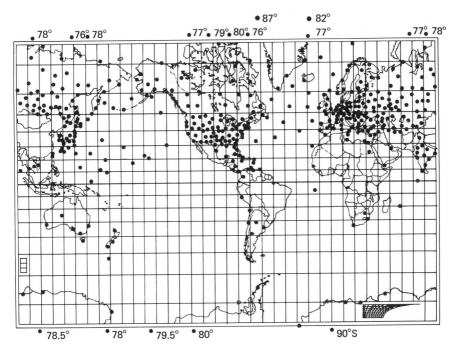

Figure 1   Global distribution of radiosonde stations (not all stations are shown in concentrated areas).

equipment (rawinsondes) to obtain upper-air winds as well as temperature and humidity measurements. There is good coverage over North America, over Europe, and even over the Soviet Union, from which we get data from regular international exchange. We also receive data from China because it is transmitted to Russia and we can monitor the transmission. There is a great contrast of coverage between the Northern Hemisphere, where data coverage over land is good, and the Southern Hemisphere and the oceanic areas where data is very sparse. In the Northern Hemisphere, however, there are significant gaps, for example, in the northern Pacific, where many trough systems form that sweep over the United States and Canada, and in the North Atlantic, which is the origin of much of the weather that hits Europe.

Dr. Hallgren emphasizes the necessity of a *new* data collection system for the world weather watch, which is more than just a patchwork extension of the present system. The United States operates about 120 radiosonde (most if not all are actually rawinsonde) stations that make two observations a day at a cost of about $100 a day for expendables: the balloons, the helium, and the radiosonde package. Neglecting labor, the United States spends about $4 million per year in expendables alone. The world has a network of approximately 500 or 600 stations. Many countries of the world use less expensive instruments and some countries have only one flight a day; but there may be anywhere from $10 to $20 million a year spent in maintaining coverage over 10 to 20 percent of the earth. Thus economically, one could afford to spend a reasonable amount of money for a system with *global* coverage. Let me emphasize that the global coverage we will be discussing will not replace the conventional networks, for it will not have the fine scale necessary for local forecasting.

Figure 2 shows the GHOST system proposed by the National Academy of Sciences,* NCAR, and the U.S. Weather Bureau, which consists of a series of inextensible balloons inflated so that they will rise in the atmo-

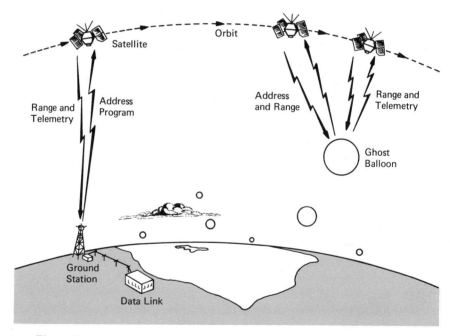

Figure 2   Schematic of GHOST balloon-satellite system showing data flow from balloons to satellite to ground station. The satellite is programmed to address balloons on a particular orbit, determine their position, and store their environmental data (temperature, humidity, ozone, and so on).

* "The Feasibility of a Global Observational and Analysis Experiment," N.S.A., Pub. #1290, 1966.

sphere to a predetermined density level and remain there for six months to two years. The weight and the size of the balloon will determine the density level at which it will have zero buoyancy and will float. The balloons are slightly superpressured; they have 10 to 25 percent excess pressure of helium in them to maintain the shape during the heating and cooling of the diurnal cycle and to provide a reserve against leakage.

The ultimate system is a series of balloons floating at perhaps 10 levels in the atmosphere. Note the comparisons of this system with the French EOLE system as described by Dr. Morel in the next article. They float along with the air parcel in which they are imbedded so that tracking them gives the wind at those levels. If they have sensors, they can measure temperature and humidity in the lower levels, and perhaps with some amount of work we can build sensors that will measure ozones in the stratosphere. Perhaps we can measure the radiation, or maybe we can deduce radiation by monitoring the superpressure of the gas.

These balloons are tracked by a satellite communication system, and each balloon has a radio receiver and transmitter. The balloon is interrogated by the satellite, which is preprogrammed so that the approximate trajectories of the balloons are predicted for any particular orbit. When the balloon hears its address called, it will transmit to the satellite all its stored data; probably these will be only the instantaneous readings of temperature, humidity, and perhaps internal pressure. The satellite ranges each balloon several times and uses these range data to calculate the position of the balloon. The satellite transmits this information once during an orbit to the ground station, where there is further processing and transmission through a data link to a central processing system.

We have made some preliminary studies of the distribution of such balloons, how the systems could work, and the number of bits required to track the balloons and send back the primitive information from the balloons. As we shall see later, the system seems to be possible within the present state of the art, but it will require a certain amount of testing first.

The development of the balloons and the lightweight electronics packages to test their performance has been an interesting program. Using mylar, which is very strong and very light, we have been able to build balloons that will reach a certain predetermined volume. Mylar, with a slight overpressure, forms a stiff structure and will retain a constant volume. As it comes out of the extrusion machines, it has millions of little pin holes, allowing gas to slowly leak out; however, if we put two sheets together, the chances of two pin holes exactly coinciding are very small, so the material is almost perfectly leakproof. The seams were another problem, but they are now taped and sealed and the manufacturers have developed techniques for making balloons that have few leaks and lose gas at nearly the theoretical rate of the permeability of the material for helium. We have tested these balloons at the NCAR headquarters at Boulder and are now preparing a flight program for the Southern Hemisphere. (See the appendix to this article for report of the test flight programs.)

The mylar balloon shown in Figure 3 has been inflated with helium at approximately 20 percent overpressure. This size balloon will float at approximately 35,000 ft. Note the taped seams. The balloon testing building at NCAR has two parts. In one the balloon is inflated to a certain lift, which is weighed off with very sensitive balances. Then it is put in the second chamber shown at the rear of Figure 3, which is maintained at a constant temperature of $40°F \pm 1°$. This is where the balloon is weighed with accuracy. Then the balloon sits, for a day, week, or a month, when it is weighed again and is brought back to the constant temperature to calculate the loss of gas.

For the test program, such a building will be used as an inflation chamber, in which a particular balloon will be taken out of a box, inflated, taken into the constant temperature room for a few days, and then weighed again. If the weight falls on the curve, the balloon is satisfactory and we launch it. If the balloon leaks, the balloon can be repaired and retested. There is no handling between inflation, testing, and launching, so that we hope to launch only balloons that have been tested and that should fly for a significant period of time—for six months at least, some in the upper levels for up to two years. If the system becomes operational in the future, manufacturing and packing techniques will have to be improved so that such tedious testing can be eliminated. The instrument package will hang just beneath the balloon.

Figure 4 shows the flight levels and the diameters of the balloons. The maximum balloon diameter is 5 m at 30 mb, and although it is feasible to go to 10 mb, the balloon becomes rather large and expensive. However, at the upper levels the balloons would have a very long lifetime, and only a few of them would be needed because the atmospheric circulation at 10 and 30 mb is more uniform than the circulation at the lower levels.

Early in the development of this system, it was thought that perhaps the best way to handle the electronic circuit was to deposit it on a thin film that would then be encased in a bag of plastic to protect it, as shown in Figure 5. Experimental circuits weighing 1 g exclusive of power supply, but including the transmitting circuits, sensors, and coding circuits have

Figure 3 Double laminated mylar balloon in NCAR testing center, sized to float at 35,000 ft inflated with helium at approximately 20 percent over-pressure.

been built. The thin-film system was designed several years ago to minimize any danger to aircraft; an airplane could hit it and hardly know it. The material had to be frangible so that it would break on impact and had to have a very low density and be spread out so that chances of it denting or penetrating an aircraft wing would be very small. The integrated circuits available today, however, should do approximately the

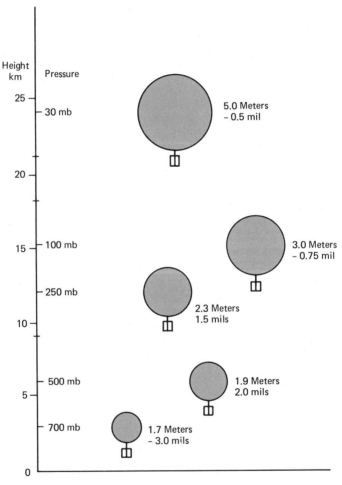

Figure 4 GHOST balloon specifications. Schematic drawing of altitude versus size of balloon required. [Based on mylar film with maximum super pressure of 15 percent; balloons weigh approximately 1 kg with the exception of 5-m balloon (1.6 kg); payload is 800 g for all balloons.]

Figure 5   Experimental thin-film balloon electronics package developed by the Schjeldahl Company.

same job, should actually be cheaper to make and easier to handle, and should not represent a hazard to aviation. This last problem quickly leads the scientist into the difficult area of politics and subjective feelings about aircraft safety.

Figure 6 is a picture of the kind of electronic package we are building now for the test flights; the future operational package will be more complex. The solar cells are provided by NASA; they are perfectly good, but they were not usable because they could not survive in the radiation belt, and so can be used only in the lower atmosphere. In the center of the solar cells is a photosensitive resistor (Figure 7), the resistance of which is a function of the temperature. The sensitive element is horizontal, and as the sun shines on it the resistance changes as a function of the sun angle. The resistance of this element then controls the repetition rate of the code-name transmitter.

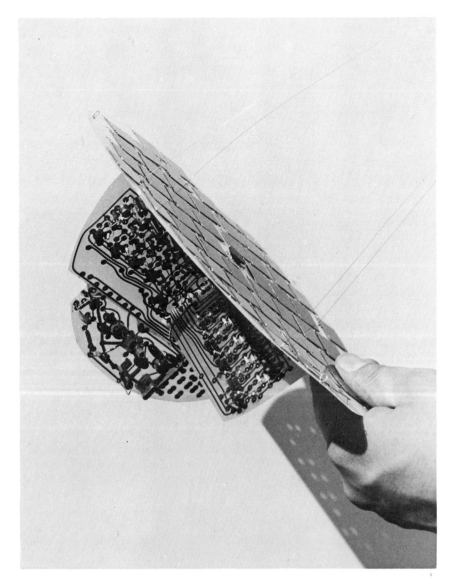

Figure 6   GHOST transmitter and encoder powered by solar cells. Photo-sensitive resistor for determining sun angle is shown in the middle of the solar cells.

The repetition rate from the balloon transmitter is carefully calibrated against sun angle so that a listener must simply tune his receiver to the transmitter, time a number of code groups, and then look up the sun angle in a table. Having done this several times during a day, the listener can make a plot of balloon sun angle versus universal time (Figure 8). The universal time corresponding to the peak of this curve gives the longitude of the balloon, and the magnitude of the peak gives its latitude (knowing the solar declination on that day). In the ultimate operational GHOST system, of course, the balloon will be ranged (or tracked in some other way) by the satellite; sun angle navigation is only intended for the presatellite test program.

This current sun navigation is sufficiently accurate to track the balloon to at least ± 100 km and may on occasion be good to ± 50 km. If one tracked the balloon once a day, then the winds are average values over the day. Since the winds are from 50 to 100 knots in the upper atmosphere, even though the tracking is relatively crude, the daily *averages* of the wind are reasonably accurate to 5 percent or so. This level of accuracy is sufficient to be interesting and useful to meteorologists.

This test system weighs about 100 g, it puts out a maximum of a

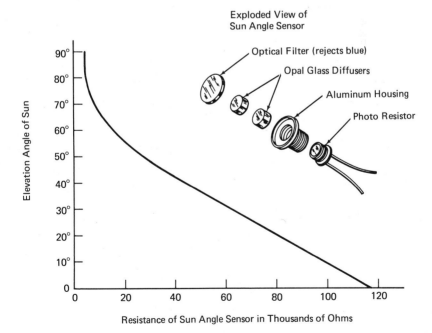

Figure 7 Sketch of sun-angle sensor.

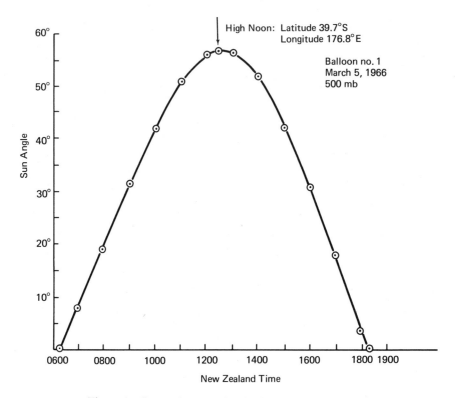

Figure 8   Sun-angle curve for GHOST balloon number 1.

little under 1 \/ RF under zenith sun conditions; it has a range of 3000 to 5000 mi; it is a relatively simple circuit; and it can provide information on four different types of sensors. The size of the package shown in Figure 6 has purposely been left larger than necessary (there has been no attempt to use integrated circuits) to keep the density of the system as low as possible.

The first step toward implementing this system is to test the balloons. In 1964, with the cooperation of the Japanese Meteorological System, we launched balloons from Japan, tracked them over the Pacific, and determined that the electronic circuits would stay in operation over that period and that we could get reasonably accurate tracking. There was trouble with the balloons in 1964, because they leaked, but in late 1964 and early 1965 the balloon was vastly improved and we made plans to track the balloons in a circumglobal flight which is now in preparation in New Zealand. (See Appendix for results of the 1966–1968 tests.)

On the nonscientific side of the program, the United States has a strong need for global weather data. The United States has the technical knowledge and the money to prepare the system, but no one country in the real world of politics can carry out a program like this. Other countries will not permit overflight of balloons unless they have previously agreed to it; and they won't agree to it unless they are sure it is useful and beneficial and that there are no dangers to their aircraft operations. It is fairly clear that most of these criteria are met. A number of balloon incidents during World War II and immediately afterward, however, have made the political climate somewhat less than optimum for initiating a program like this. Therefore, it was decided that world cooperation through some established international channel was essential for such a system.

One very good semigovernmental channel is the World Meteorological Organization (WMO), which is a special agency of the UN. The members of the WMO come from the weather bureaus of some 126 countries. The WMO is not an operational organization, but it is a very good forum for discussing technical and political problems in connection with meteorological activities. (See the article by Dr. Malone in Part 2.) Through the WMO, countries have agreed to standardize observations, to transmit their data at regular times and with regular codes; we get the Chinese data, for example, because they use the WMO code. At the moment, the WMO seems to be the best channel in which to attempt to achieve international cooperation.

The GHOST balloon system has been discussed by the WMO, and in the spring of 1965, the Executive Committee agreed that in principle this was a feasible solution to the problem of obtaining world weather data. The Soviet Union agreed, for the first time anywhere, that a test of such a system would be acceptable, providing that the WMO work out the rules under which the test would be carried out and providing that the first test be carried out in the Southern Hemisphere. This, however, was no detriment to the program, because it is exactly the Southern Hemisphere where we need the data the most.

The French also have made a careful study, and now that they are a satellite-launching nation are also in the position unilaterally to mount a global weather program. (See the next article, by Dr. Morel.) A few years ago the French asked the United States for their cooperation. Since then several groups in the United States have been in close contact with the French in exchanging ideas and plans. The French plan is a little different from ours; it does not depend on programming the satellite for serial interrogation of the balloons. The French have decided to limit their first operational test to about 500 balloons in the tropics. The ultimate system will include perhaps 5 to 10,000 balloons at 6, 8, or 10 levels in the atmosphere.

We have gone through in a very preliminary way a study of some of the technology of the system as it might develop and it appears to be quite feasible using present observing techniques and instruments. Each balloon might transmit about 100 bits of information to the satellite for its internal measurements; this could include its position as computed in the satellite. The satellite would range the balloon and compute its position based on programmed information supplied to the satellite from the ground station. Perhaps on one path, a polar satellite could intercept as many as 2000 balloons. It would intercept almost all the balloons in the polar areas and only a fraction of the balloons in the equatorial areas. This would be a maximum of about 200,000 bits for a satellite to transmit to the ground control station for a particular orbit during a period of 10 sec up to maybe a minute, depending on propagation conditions. But this is sufficient time to transmit 200,000 bits, and even if the system were to require a 10 to 1 redundancy and have 2 million bits, there would still be fewer bits than are required to transmit one of the many high-resolution Tyros or Nimbus pictures. This is comfortably within the present satellite capabilities. If one wished to add more parameters to the balloon, perhaps the bit rate would come close to that presently achievable, but it looks as if this is not a serious problem.

The balloon problem itself is nearly solved, at least for balloons to operate at 200-mb levels and above. We are quite sure that a reasonable package operating in the UHF band can be built by the use of some of the new solid-state devices that could operate at a minimum power of 10 to 100 mW. Transmission from the balloon to the satellite ought to be much simpler than the transmission from the satellite to the ground.

The one part of the system that is the least developed is the power supply; solar cells are fine, but they get only half-a-day operation in the equatorial regions and very limited operation in the winter polar nights. However, initial developments have been made in thin-film batteries, and the Diamond Ordinance Fuze Laboratory in Washington has built a single experimental thin-film battery that will withstand many cyclings and will operate at $-60°C$. It remains to be seen whether a practical thin-film battery can be manufactured.* It seems likely that with some reasonable effort it will be possible to build a combination solar cell battery storage system that will provide for nighttime and polar winter operations. Even a daytime-only system would be a significant advance toward global coverage.

* The French have succeeded in developing a practical thin-film battery that is useful down to-50°C.

To test for hazards, some components of the system have been thrown into a jet engine. They came out in tiny pieces without doing any operational structural damage to the engine.† Some of the earlier packages were fired into an aircraft wing at a velocity simulating a jet velocity of 600 mi/hr and have broken the skin, so that some work needs to be done on reducing the weight and keeping the density low enough that if there is impact, the package will simply shatter or bounce off. The balloon itself should not be a hazard.

These safety problems are hard to theorize about since the aviation authorities of many countries are very nervous about balloons flying around at aircraft operating levels. Radiosondes weighing several pounds are enough worry; but it is not quite legitimate to say that because people tolerate radiosondes they should tolerate balloons. We tolerate radiosondes because we have to have them (considerable pressure has already been exerted on the meteorological community to develop thin-film or lightweight radiosondes). The radiosonde, however, flies up and down through the air lanes, but the GHOST balloons will float in the air lanes permanently. Because it is crucial to get the weather information at the lower levels, where there is a great deal of air traffic, a considerable amount of concern is expressed that the design of the system be completely safe and nonhazardous under all conditions. We are quite sure that with the work that has been done so far and with the tests that have been made, the GHOST system will be safe and that the result will be acceptable to the aircraft authorities.

In order to achieve maximum cooperation, besides having the usual number of meetings with representatives of the different agencies, we have appointed one of our staff scientists at the invitation of NASA to serve as the project leader for the design phase of a satellite balloon system. He will take up residence at the Goddard Space Flight Laboratories near Washington and will be part of the design group. Hopefully this will stimulate a healthy level of communication between the meteorologists who want the system and the engineers who have to design it. NASA has many capable engineers who could design new instruments for almost any system at any time, and they feel a strong need to tailor their engineering work and their ultimate system to the real needs of the scientific and operational community. This appears to be one of the very hopeful signs that ultimately, in five years or so, this system will actually come into being.

---

† Design criteria have been developed in France and the United States for the Construction of nonhazardous packages.

APPENDIX

STATUS REPORT ON GHOST TEST FLIGHTS
IN THE SOUTHERN HEMISPHERE NOVEMBER 1968

Flights began on 4 March 1966, from Christchurch, New Zealand, with balloons designed to float at 500 mb. Several flights achieved durations of 15 to 20 days, but as winter weather set in and the ballons encountered cold fronts and icing conditions, failures shortened the flight times. It was deduced that water was condensing on the balloons and then freezing, adding enough weight to cause the balloon to descend and eventually lose overpressure. Later, tests were tried with waxed balloons, and this treatment drastically reduced the water load that the balloon picked up. However, this kind of treatment proved to be short-lived. Investigations are now underway with manufacturers to develop a more permanent hydrophobic surface treatment.

In late March, flights of ballons designed to float at 200 mb commenced. The first such balloon, LLL, made many circumpolar trips before it was finally lost after some 76 days (Figure 9). A significant portion of the 200-mb balloons had lives of over 100 days and several flew nearly one year. A balloon of 100 mb is the record holder; when last heard (6 November 1968) it had been in the air 405 days and is expected to stay aloft longer.

Round-the-world flghts are averaging some 10 to 15 days. Occasionally balloons seem to get caught up in the antarctic circulation, sometimes for many days, and then reappear in the mid-latitude circulation. We lose track of these balloons south of about 60° because of low sun angle.

The 200-mb flights are considered a success, even though they are not yet approaching their theoretical lifetime of 600 days. They were not tested exhaustively prior to launch, however, owing to difficulties arising from their large size and the need to conserve helium at the launch site. It is felt that refinements of manufacturing techniques and packaging methods for shipment will provide a vehicle that will have a satisfactory performance.

The tracking techniques has been an unqualified success. Figure 9 shows the remarkable uniform tracking curve for a 200-mb balloon. Tracking of the 200-mb balloon has been achieved around the world with only two stations

250

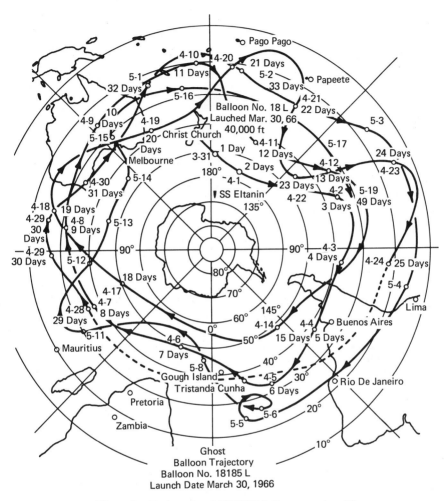

Figure 9   Trajectory of GHOST balloon number 18.

on many occasions, giving propagation paths of more than 10,000 km. a computer program is under development that will give a best fit of tracking data and provide balloon trajectories and wind velocity data hitherto unavailable in the Southern Hemisphere. Table 1 summarizes U.S. and French balloon tests as of November 1968.

A GHOST package was also fitted with a battery supply and mounted on a mast at the Christchurch site. It telemetered electronics temperature and was successfully monitored continuously in Melbourne. Thus, the GHOST

TABLE 1
BALLOON PERFORMANCE

Level	Number	Average Life	Maximum Life	Outlook for Average Life	Failure Mode
30 mb	10 U.S.	50 days	116 days	$\approx$ 1 year	Ascent damage
100 mb	22 U.S. 10 French	>100 days	>400 days	$\approx$ 1 year	"
200 mb	60 U.S. 20 French	90 days	$\approx$ 1 year	170–200 days except in tropics	Pinholes; icing in cirrus
300 mb	50 French 5 U.S.	12 days 43 days	102 days 88 days	50 days but 15 in tropics	"
		—	—	—	Icing
500 mb	20 U.S.	7 days	22 days	—	Icing
700 mb	3 U.S.	14 days	21 days	20 days	Diffusion loss; icing at mid-latitudes; possible mountains.
850 mb	10 French	11 days	20 days	20 days	Diffusion loss; mountains

concept may also provide a basis for inexpensive remote surface weather stations, especially suitable for the island chains in the Pacific.

PIERRE MOREL

Pierre Morel is an Associate Professor at the University of Paris. He is the scientific director of the French EOLE project, which is a global free-floating balloon data system. Dr. Morel directed the early design of the EOLE system, and he is now in charge of the meteorological aspects of the project.

Abstract—*Dr. Morel follows Mr. Ruttenberg's discussion of the U.S. GHOST system by describing the French EOLE satellite balloon data network.*

There will be two EOLE experiments (EOLE A and B), both of which will consist of one tracking satellite with approximately 500 balloons floating at the 300-mb level in the atmosphere. Dr. Morel outlines the general purposes of a balloon data network and describes the satellite and its electronic package in some detail. He also discusses the data requirements both on the satellite and for the individual balloons.

The EOLE Experiment

THE first problem for any meteorological system is that of obtaining a general idea of the motion of the atmosphere about the earth. This is much like the desire of the man working in fluid mechanics to see the motion of his fluid. The best way to do this is to put tracers into the fluid, whether this is the atmosphere or water. In the atmosphere, since the fluid is very large, the tracers to be used in the immediate future will be balloons moving at constant levels. A numerical model for computing or simulating the motion of the atmosphere might use a grid system 1000 km long, 1000 km wide, and 10 km deep. In each box of the grid there would be one, two, three, or in the most complete system, perhaps ten different levels. Within each box would be ten different grid points, at which initial values of the temperature, pressure, and water-vapor content must be supplied to the model. The exact nature of the data required depends, of course, on the variable being used in the model. This problem has been discussed earlier by Dr. Leith and Dr. Mintz in Part 6.

About 250 boxes would cover the Southern Hemisphere with such a grid system—a volume of 250×10^6 km^3. Now, if we spread balloons randomly about the Southern Hemisphere, in order to be reasonably certain of having at least one balloon in each box, we need about twice as many balloons as we have boxes, or about 500. This is the smallest number of balloons that will provide a reasonable distribution throughout a large area such as the Southern Hemisphere. On the other hand, this is a bare minimum because these 500 balloons would cover only one level. A six-layer model (such as that of Dr. Leith) would require 500 balloons at six separate levels. In addition, if one wished to have a box that was, say, 500 km on a side, instead of 1000 km, then one would need four times as many balloons. So, in total, we would need $4 \times 6 \times 500$ or 12,000 balloons for a complete six-level meteorological system in the Southern Hemisphere. The numbers needed then are between the bare minimum of 500 balloons and the maximum of 12,000 balloons.

255

Dr. Leith uses a variable square mesh of 5° (as described in the appendix to his article in Part 6), which means that there are approximately 1000 grid points in the Southern Hemisphere. Thus the number of data balloons required by Dr. Leith's six-level model is 6000. Dr. Mintz, on the other hand, uses a 9° × 7° grid plus polar caps, which means there are 500 grid points over the Southern Hemisphere. Thus for his two-level model, 1000 balloons would be required.

The EOLE experiment consists in setting up and maintaining for a limited period of time a network of 500 instrumented constant-level balloons drifting in the atmosphere and tracked by a special-purpose interrogation and communication satellite. The EOLE system is designed for direct (in situ) measurements of the fundamental atmospheric parameters: wind, velocity, pressure, temperature, and eventually, relative humidity.

The wind sensors are the platforms themselves since the so-called constant-level balloons drift with the wind on a constant density surface and act as tracers of the air masses. A significant amount of experimental and theoretical work has been done and described in the literature exploring the significance of this horizontal sounding technique. In the EOLE experiment, each balloon will be interrogated twice a day on a two successive daylight passes of the satellite, that is, at a 100-min time interval. From these interrogations two successive locations of any given balloon will be computed, thereby providing, once a day:

(1) one measurement of the wind averaged over a 100-min period;
(2) one fix on the trajectory of the atmospheric masses since the previous day.

It now seems certain that on the basis of preliminary laboratory tests of the radio equipment and statistical computations of the geometrical factors involved in the localization procedure, any one fix given by the system should fall within a 5-km error cycle.

Temperature sensing is straightforward in the troposphere. Although constant-level balloons will not provide ventilation, commercially available thermistors will give free air temperature within 0.5°C.

Lightweight, frictionless pressure sensors are not currently available, but are nevertheless well within the present state of the art. A miniature capsule with variable inductance pickup has been developed for the preliminary EOLE flights. The weight of this capsule is 3.5 g. The capsule has been used to measure the differential pressure inside the balloon with an accuracy of ±1 mb in the range 0 to 60 mb.

A dependable miniature humidity sensor suitable for EOLE has not been found yet. The technique of making metallic oxide elements has progressed significantly during the last years. It is not known, however, whether their characteristics will be stable enough for drift-free operation during an extended period.

Some preliminary work has been started on a possible extension of the EOLE sensing system to drifting buoys. The parameters which could be measured from a buoy include:

(1) surface wind speed,
(2) locate this platform with sufficient accuracy;
(3) water temperature,
(4) sea state,
(5) relative humidity at different heights above the sea surface, and
(6) cloudiness.

The EOLE system comprises one satellite package and a large number of transponders placed on suitable platforms like balloon gondolas, buoys, or automatic ground stations. This system must

(1) interrogate one particular platform among all those present within the visibility radius of the satellite (of the order of 3000 km in the present system);
(2) locate this platform with sufficient accuracy;
(3) transfer meteorological data acquired by the platform instrumentation into the satellite storage recorder;
(4) allow for multiple readout of the content of this recorder by suitable ground stations without destroying the stored data.

In addition, the system must fulfill several important design goals:

(1) The balloon-borne equipment, including transponder, sensors, and power supply, must be significantly lighter than conventional radiosondes.
(2) The interrogation-location-telemetry operation must be as short as possible to allow the interrogation of many platforms which may be concentrated in the visibility range.

Since direction finding from a satellite is quite difficult, the location methods are restricted to range or range rate measurements. A range determination, based on the measurement of the two-way transit time of a radio signal from the satellite to the balloon, will locate this balloon on a certain sphere centered at the satellite. Alternatively, a range rate measurement can be derived from the Doppler effect of the returning signal; it will place the answering balloon essentially on a cone with its axis tangent to the satellite trajectory. Any two measurements of the first or second kind will give a fix of the answering balloon. Both approaches have been studied, up to the electronic breadboard stage, and finally the range rate system has been selected.

This system is based on a two-way coherent radio link between the satellite and the balloon. The satellite transmits continuously a stable carrier frequency (460 MHz) and binary addresses. Upon recognition of the proper address, the interrogated balloon transmits back a rational multiple (such as $\frac{7}{8}$) of the incoming signal frequency. The balloon transponder thus maintains phase coherence between the incoming radio signal from the satellite and the radio wave transmitted back to the satellite. The Doppler effect due to the relative radial velocity of the balloon with respect to the satellite can be measured with excellent accuracy by comparing the signal from the balloon and the satellite clock frequency.

A special phase-lock receiver capable of acquiring the signal in less than 100 msec has been developed. The feasibility of the entire system has been demonstrated by actually locking onto the signal from Transit satellites measuring the Doppler shift and computing the location of the (ground-based) experimental receiver.

Without going further into the design of the electronic system, computations and experiments show that a 1-to-5-km location accuracy can be achieved while keeping both satellite and balloon radio power to a reasonable 2-W level.

Two successive interrogations of a given platform are needed to obtain a fix, that is, two range or two range-rate measurements. These interrogations could very well be programmed into the spacecraft memory by ground command; the interrogation signals would then be transmitted when, and only when, the platform is most likely to be in view of the satellite. This would somewhat complicate the operation of the satellite but still be perfectly acceptable for an operational system (it would in fact be necessary if the number of platforms is to exceed 500 or 1000).

For the EOLE experiment, however, it was thought best to go around this complication by calling cyclically all 512 possible addresses all the time. Also, the cyclical nature of the address sequence has allowed substantial simplifications of the balloon digital circuits. Since calling a particular platform and recording its answer lasts only 0.25 sec, the complete sequence of 512 calls lasts about 2 min, allowing ample time for at least three successive interrogations of all transponders within a 2000-km-wide strip on each side of the earth track of the satellite.

The system allows the transmission of 4 digital words of 8 bits on each interrogation of any particular balloon, that is, 12 to 24 words during a satellite overpass. Actually, 3 words only will be used in the EOLE-A experiment to carry the following data:

(1) free air temperature—8 bits,
(2) local pressure—8 bits, and
(3) overpressure of the balloon—8 bits.

To this must be added the following data generated or measured by the spacecraft:

(1) time of the interrogation—5 bits,
(2) balloon address number—9 bits, and
(3) range or range rate—11 bits.

Each successful interrogation will therefore account for about 50 bits of information. The EOLE satellite will be provided with a 100,000-bit digital memory large enough to store the results of four successful interrogations of each of the 512 balloons. Observe that such digital memories are well within the present state of the art.

The satellite will also have memory space to store command orders addressed to selected platforms. These orders will be sent from ground stations to the satellite and broadcast thereafter each time the selected addresses appear in the interrogation sequence. This command subsystem provides a communication link from the ground control facility to any particular platform anywhere on the earth via the satellite memory and interrogation transmitter. The EOLE satellite will therefore act as a courier carrying orders from the control center to selected platforms and bringing back to the center the meteorological data acquired by all platforms. In particular, this command link will be used to destroy straying balloons, remotely.

The EOLE satellite will be a small spacecraft (60 kg approximately) designed and built for this particular experiment. Original features of the spacecraft will be:

(1) gravity-gradient passive attitude control along the local vertical,
(2) highly stable quartz oscillator (developed for the D-1 geodetic satellite program) used as main clock for all satellite functions,
(3) random access magnetic core memory, and
(4) directive 400-to-460-MHz antenna for satellite-to-balloon communications.

All balloons will be distributed around the Southern Hemisphere at one level only (300-mb nominal). Each craft will consist of a leak-free spherical pressurized balloon carrying the instrument gondola and a flotation device to prevent the loss of the craft in case of a severe icing, which could overload the balloon and bring it down to the sea.

Superpressure balloon envelopes have been developed in both the United States and France. For medium altitude (300 mb) the envelope is made of 16 to 24 gores assembled with adhesive tape. The fabric is 50 μ bilaminated terphane (mylar). When pressurized, envelopes assume a near-perfect spherical shape. Balloons with 2-m and 2.5-m diameters have been fabricated and successfully tested in flight under nominal overpressures of 60 and 40 mb, respectively. No measurable loss of lifting gas was found during a 40-day flight, thus showing that leak-free envelopes can be obtained and that diffusion of helium through the fabric is negligible.

The gondola includes a directive antenna, the 400-to-460-MHz coherent transponder and associated addressing circuits, meteorological sensors, a solar cell generator, and a power storage battery. This equipment will be comparable in weight to the present-day radiosondes but dispersed in a much larger volume so as to prevent damage should a collision between an aircraft and balloon occur. Simulated collision tests are currently being conducted by shooting electronic equipment at aircraft structures.

The balloon equipment will be subjected to the low ambient temperature ($-55°$ to $-30°C$). This presents no difficulty for transistors and passive electronic components. On the other hand, no power storage battery works well at this low temperature. It has been demonstrated, however, that a single nickel-cadmium element on a thin substrate can deliver the minimal amount of power needed to operate the balloon receiver in stand-by conditions and to provide 10 W peak power during short pulses, even at night.

Preliminary flights have shown that the most likely cause for balloon destruction is icing of the envelope. Such icing occurs either at night if the balloon temperature is lower than the frost point of the ambient atmosphere, or during the day when the balloon enters thick ice clouds. In both cases, icing goes on until the free lift decreases to zero, whereupon the balloon starts its descent toward ground or sea level.

No device has been found yet to prevent the deposition of ice on the envelope. One can prevent the loss of the craft after such an icing by adding a float acting as guide rope when the craft hits the sea. The gondola will then stay high above the sea until the ice melts away and the balloon lifts back up again. This scheme has been tested successfully with 400-mb balloons.

The regularity of the mid-latitude westerlies makes it possible to distribute the balloons evenly from as few as five launching stations set 1000 km apart on a meridional line extending ideally from 20° to 60° south.

However, balloon trajectories in the tropical zone are bound to be much more complicated. Several additional launching stations will therefore be needed around the tropical zone.

The launching schedule from the mid-latitude stations will be rapid enough to make up five streak lines of tracers in the main circulation during the first two weeks of the EOLE-A experiment.

The EOLE location system is basically restricted to relatively low-altitude satellites; the higher the spacecraft, the larger the error. The orbit must therefore be chosen as low as possible to improve accuracy and yet high enough to provide an adequate coverage or visibility range. A circular 1000-km orbit would be the best compromise. Very distant orbits (like the synchronous orbits) must definitely be excluded for such applications.

Since quick recovery of all data stored in the satellite is desirable, it would be advantageous to choose either a polar or an equatorial orbit. A polar satellite will pass in view of a high-altitude station once every orbit. Similarly, an equatorial orbit will bring the satellite over an equatorial site once per orbit.

As a matter of fact, the orbit chosen for the first EOLE satellite is neither polar nor equatorial, but has a 50° inclination on the equator; this particular orbit maximizes the coverage of the mid-latitude zone, which is the locus of this experiment.

Two separate experiments (EOLE A and B) are considered in this program. Before these, preliminary constant-level flights will be conducted in order to test the balloon equipment and to simulate a series of launches from one station. The preliminary flights use a HF radio signal derived from the GHOST beacon. Results obtained are similar to GHOST with the addition of overpressure and/or temperature measurements.

EOLE-A is primarily directed toward the study of the mid-latitude circulation around the Southern Hemisphere. The altitude chosen (300 mb) is close to the jet stream level and is well suited for observations of the upper-troposphere winds. In the mid-latitude where the flow is not far from the barotropic flow, single-level wind fields will provide sufficient information on the synoptic situation. This experiment is planned for 1969.

EOLE-B will be devoted to observing the tropical circulation. Since this flow has a definite cellular pattern, several levels of tracers will be needed. These levels are not defined yet, but on the basis of preliminary flights one would suggest the following:

(1) 850 mb, lower troposphere,
(2) 300 mb, upper troposphere, and
(3) 100 mb, stratosphere.

The EOLE-B experiment will be restricted to a limited area ranging from 25°N to 25°S and extending over about one-fourth of the equatorial belt. This restricted horizontal coverage will allow a denser network of tracers and several levels. Study is underway to include automatic drifting buoys in this second experiment, which is planned for late 1970.

Since the EOLE satellites are definitely not operational, no special effort is planned to provide really fast data acquisition and processing. The orbit of EOLE-A, for example, has been chosen for the best coverage but not for the fastest data transmission (a polar orbit would have been more adequate). Generally, the satellite will pass over any of the four ground stations of the French space data acquisition network (IRIS) located at:

(1) Canary Islands,
(2) Ouagadougou (Republic of Upper Volta),
(3) Brazzaville (Republic of the Congo), or
(4) Pretoria (Union of South Africa).

On command from these stations, the satellite will transmit the data stored in its memory and possibly receive further operational orders. These data (100,000 bits) will be temporarily stored at the ground station and thereafter transmitted to the *Centre Spatial de Bretigny* via commercial telex lines. The data will be immediately fed to a computer and reduced to a readily usable format specifying the location of each interrogated balloon. Hopefully, the maximum delay between data acquisition and the presentation of reduced data will not exceed six hours. The reduced data will be disseminated through usual telex links.

The reduced data format is:

(1) platform number (address),
(2) time of interrogation,
(3) location,
(4) local pressure and temperature, and
(5) status (including overpressure in the balloon).

The operational schedule (destruction of a particular balloon or launching of new balloons to fill gaps) will be issued immediately after production of these reduced data. It must be stressed that the satellite interrogation subsystem will provide status information about all live balloons including those being readied for launching at field stations. The satellite is itself the most efficient test set for checking out the balloon payloads before launching. Actually, the Bretigny control center will know better and sooner than field operators the status of each live payload and can easily order the launching of any particular balloon using the return link via the data acquisition stations and the spacecraft itself.

Analysis of radio communication over the satellite to balloon distance shows that the accuracy of range or range rate measurement will yield an 0.8-km error. The location of a balloon with respect to the satellite track is determined by three range measurements, and the overall accuracy of the fix will be of the order of 3 km. This uncertainty is to be compounded with those due to the satellite position error and to the effect of random wind. Present-day satellite tracking techniques are such that the position of the spacecraft can be known at all times with a probable error less than 100 m. For the present purpose, the satellite trajectory can, therefore, be assumed to be perfectly determined. On the other hand, random winds (gusts) do introduce a sizable and quite irreducible location error of the order of 1 km per 10 km/hr of random wind speed. The final location error can, therefore, be estimated to be of the order of 5 km or less on each individual fix.

The interrogation location process lasts approximately 256 msec. This time places a limit on the number of balloons which can be successfully interrogated within a given visibility range. Assuming a minimum mutual visibility time of 7 min on the edges of the area overflown by the satellite, about 500 balloons at most can be interrogated three times during the pass.

A further limitation is imposed by the very simple interrogation logic contemplated for the EOLE experiment. Since the 512 addresses are called cyclically, one cannot operate more than 1000 or 1500 stations altogether at a given instant (two or three stations answering to the same address number would not bring any difficulty provided these stations are sufficiently far apart).

The telemetry channel performance is quite low in the EOLE experiment project since we have deliberately chosen to transmit only three words of 8 bits to measure only the local temperature and pressure. The data-handling capability of the balloon to satellite communication link is, however, much larger. As many as 600 bits of information could easily be accommodated on this link with optimum digital coding (at the price, however, of providing a more sophisticated data encoder on the balloon and a larger storage memory on the satellite).

MYRON LIGDA

Myron Ligda received a B.A. degree in 1942 from the University of California at Berkeley in astronomy and physics. He studied meteorology at New York University and the Massachusetts Institute of Technology, receiving an M.S. degree in 1948 and a Ph.D. degree in 1953. From 1954 until 1958, Dr. Ligda was Associate Professor on the faculty at Texas A and M and was the principal investigator of the Radar Meteorological Section of the Department of Oceanography and Meteorology. From 1958 until his death in October 1967, Dr. Ligda was at the Stanford Research Institute as the Director of Research Programs on Atmospheric Analysis, and Radar Satellite Meteorology. He was particularly interested in the use of radar and lidar (laser radar) for weather observation.

Abstract—*The lidar, which is the optical equivalent of radar, even though it is still in its infancy, shows great promise of becoming a very useful tool for atmospheric measurements, both from the ground and from satellites. Dr. Ligda discusses several possible uses for the lidar, some of which, such as elementary cloud height measurement, are currently possible. Dr. Ligda is careful to point out, however, that other uses, attractive though they may be to the meteorologist, will have to wait for further developments in the lidar itself.*

The Use of Lidar
for Weather Observation*

LIDAR or laser radar, an instrument now under intensive development at the Stanford Research Institute (SRI), has a number of interesting meteorological applications. Lidar is basically a radarlike apparatus which measures the intensity of the radiation backscattered from a ruby laser beam. Since there is a direct correlation between the time at which the backscattered radiation returns to the source and the distance the beam has traveled, it is possible to determine the height and density of cloud layers directly from a plot of time vs. intensity of backscattered radiation.

Lidar is currently being used in controlled laboratory conditions to measure smoke trails and the distances of clouds above the earth. However, given substantial development effort, there seems to be no serious reason why it could not be used to make cloud height measurements of even low-density cirrus clouds at night from a 1000-to-1500-km satellite using a radiated energy from the ruby laser of 1 J per sounding and a receiving aperture of 1 m². Denser clouds at lower altitudes would be more easily detected by the suggested system.

The following paper is instructive in that it surveys the numerous factors that must be taken into consideration in very early stages of the design of a new and novel weather satellite. As will be noted, many tradeoffs and compromises must be made. Some of these are controlled by rigid physical facts. Some are strongly influenced by financial and technological considerations. A few are dictated by the scientific requirements. In his interest in, and enthusiasm for, the system he is designing the engineer should continually keep in mind the needs of those who will utilize the

* This article is drawn from a lecture given at Stanford by Dr. Ligda on 20 January 1966 and from the SRI report by W. E. Evans, E. J. Wiegman, W. Viezee, and M. G. H. Ligda, *Performance Specifications for a Meteorological Satellite Lidar* (prepared for National Aeronautics and Space Administration). Menlo Park, Calif.: Stanford Research Institute, June 1966.

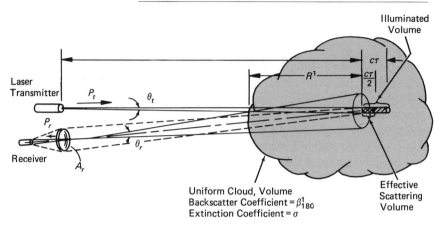

Figure 1   Schematic drawing of Lidar system.

data and, if possible, consult widely with knowledgable representatives of the user community.

A lidar, as shown in Figure 1, consists of a laser which serves as a powerful, monochromatic light source; and optical system for shaping the light into a narrow, concentrated beam; an optical system for collection of the backscattered light; a filter for the exclusion of light of undesired wavelengths; and a photodetector for conversion of the received light into electrical signals. In a pulsed lidar system, the laser is made to emit its energy in extremely brief bursts, perhaps as short as 30 to 50 nsec in duration. As this light pulse travels away from the lidar, matter in its path scatters and absorbs it. That portion of the light scattered backward toward the lidar may be detected by the receiver section. The distance to the matter scattering the energy then may be readily determined by measurement of the time interval between the transmitted pulse and received "light echo." The direction to the scattering matter is determined from knowledge of the direction in which the transmitter (and, of course, the receiver) was aimed. Thus the position of the matter relative to the lidar is established.

A laser makes a preferred light source for a lidar system for a number of reasons. First, the light it produces is extremely monochromatic; this makes it possible to employ extremely narrow bandpass filters in the receiver and thus minimize undesired background (white) light from such powerful sources as the sun. Second, the light a laser emits is coherent, or in phase, enabling the use of relatively small-aperture optical systems with sharp-beam focusing to conserve energy. Third, in Q-switched laser systems extremely powerful, brief pulses of light are produced which result in greater detection ranges and better range resolution for the system.

A Q-switched laser is one which stores a very large amount of energy prior to discharge (during brief portions of the operating time more optical power is emitted than electrical power applied) by controlling the Q of the laser cavity to delay the time development of the pulse.

A typical lidar system might employ a Q-switched ruby laser producing pulses of light containing half a joule of energy over a period of 40 nsec for a peak power of about $12\frac{1}{2}$ MW, a transmitting aperture of perhaps 6 in., giving a beam width of about 1 min of arc, a receiving aperture of roughly similar size to that of the transmitter, a receiver filter with a bandpass of a few angstrom units, and a highly sensitive photodetector. A number of such systems have been assembled by SRI for experimental meteorological observations, and much of the information contained in this article is based upon experience gained with these systems.

The list of known atmospheric constituents potentially detectable by a lidar-equipped meteorological satellite is impressive enough. We will here enumerate the wide range of possible observations, but will give a minimum of attention to technical feasibility, since many of the phenomena mentioned below could not be detected with the lidar system proposed by SRI, which has as the upper limit of its capability the bare detection of the Rayleigh backscatter from the atmosphere at an altitude of 10 km. Until surface-based lidar observations have provided more information about the occurrence and optical properties of some of these phenomena, their measurement from satellite altitudes must necessarily be highly speculative.

Starting at the top of the atmosphere, the first major particulate constituents to be encountered are the meteoric dust layers or trails. Micron- and submicron-sized particles drifting in from outer space or left behind as debris from the downward plunge of larger bodies are plentiful, especially on the forward hemisphere of the earth as it moves in orbit around the sun. The ability of lidar to detect meteoric dust seems possible if Fiocco's observations (1963)* are a reliable criterion. The global observation of meteoric dust on a systematic basis may allow more detailed examination of possible relationships between precipitation and the earth's encounter with, or attraction of, extraterrestrial matter.

In this same region of the upper atmosphere the lidar may detect the elusive *leuchtstreifen*. Again, Fiocco has reported observation of particulate matter concentration in this region (90 to 180 km). Observation

---

* References for all the observations and techniques reported in this article are given in the original SRI report.

of these "clouds" is evidently so rare that their very existence is somewhat problematical, but surface-based lidar observations may clear up this question in a few years. Of course, at this level and below, lidar might fortuitously detect the exhaust plumes of rocket engines and the trails of reentering artificial satellites.

At somewhat lower levels (80 to 90 km) the noctilucent clouds present themselves as very interesting targets. If these can be detected by satellite-borne lidar, many questions concerning their global distribution, diurnal variation, and vertical structure, and possibly their nature and origin, may conceivably be resolved. Efforts to detect these clouds with surface-based lidar have not yet been successful (so far as we have been able to determine); but it seems not unreasonable to expect that their observation only awaits the assembly of suitable equipment and its operation while the clouds are within range.

In a similar category with noctilucent clouds but in the 20-to-30-km layer are the rare nacreous, or mother-of-pearl, clouds. Much the same type of information on these clouds may be obtained from their observation by lidar as in the case of the noctilucent clouds. SRI scientists have reported the observation of what definitely appear to be particulate layers in this region.

In this region of the upper atmosphere are also to be found the maximum ozone concentration and Junge's 22-km ammonium sulfate layer. The latter has been tentatively detected by surface-based lidar, and the former *may* be detectable by its attenuating effects on high-powered infrared or ultraviolet lidar systems. Also in this region are the cirrus blowoffs from especially violent thunderstorms and dust from major volcanic and H-bomb surface explosions, all of which are promising phenomena for observation. Considerable interest is currently centered on the circulation of the atmosphere in this region, such as the vertical-transport rates and processes affecting the transport of aerosols and ozone from above to below the critical "Junge layer" near 22 km.

The highly reflective properties of most of the usual cloud types of the lower, middle, and upper troposphere make them excellent lidar targets, about which more will be said later. Precipitation, although quite reflective to laser beams, will normally occur beneath rather thick and opaque cloud formations and so should only rarely present a detectable target. An exception to this may be snow generated from thin cirrus. Contrails should be readily detectable, but whether observations can be made at short enough intervals to distinguish them from thin cirrus layers and trails is open to question, at least in early-generation systems.

While requiring laser power levels, which present the possibility of causing eye damage to persons on the surface who might accidentally look straight up the beam toward the lidar satellite, interesting possibilities exist for making worthwhile observations of atmospheric conditions in the cloud-free regions of the troposphere by means of backscattering from the relatively low-density but all-pervasive particulate matter. Such phenomena as dust storms which were apparent in the TIROS VII observations over the Persian Gulf on 11 April 1964 (similar to those shown in Part 5 by Dr. Schnapf), haze and smoke layers, volcanic dust, and regions of blowing snow all appear to be observable with a suitably designed lidar satellite system. It has been suggested that air masses and the boundaries between them may be distinguished by lidar, since a "sea-breeze" front has apparently been observed with a surface-based lidar by the aerosol discontinuity across it. Those temperature inversions accompanied by smoke, haze, and dust variations are also within the realm of possibility of detection. Inasmuch as the tropopause is often coincident with the boundary between aerosol-laden tropospheric air and the less turbid air of the stratosphere, satellite lidar observations might occasionally detect the height of this important region of the atmosphere.

The foregoing summarizes those phenomena for which there seems to be at least a slight hope of observation with a satellite-borne pulse lidar system of adequate power and receiver sensitivity. It has been suggested that additional atmospheric phenomena might also be detectable by various optical systems incorporating lasers and more sophisticated signal-analysis techniques. We shall briefly discuss several of these and note the special technological and practical problems each evidently presents.

There are two suggested ways in which lidar systems might be able to detect turbulent regions in cloud-free air. The first is based upon the use of the Doppler shift of the backscattered return somewhat along the approach used with Doppler radar to determine the component of raindrop velocity along the beam. The other is based upon the way that radar detects turbulent regions in thunderstorms, namely, by distinctive patterns or characteristics of the precipitation echoes. There has been hope that lidar might operate in a similar fashion from the aerosol return always present to some degree even in very clean air.

Even at very short range, no positive observational evidence with lidars has yet been obtained that either of these hypotheses is correct despite the considerable experimental effort expended so far in their evaluation. Also, there is as yet no proof from other types of observations that significant aerosol or density gradients are present in turbulent regions of the upper or lower atmosphere. A satellite lidar system for CAT (Clear Air Turbulence) detection would evidently need to observe over large areas with exceedingly good horizontal and vertical resolution to provide useful information.

Following the approach of scientists who have employed searchlight beams to measure upper-atmosphere density by Rayleigh molecular backscatter, experiments have been performed with lidars toward the same goal. To date, these have not been highly successful because of the difficulty of measuring the intensity of the very weak returns obtained, although the potentialities are interesting. There is a need for much more power and a shorter interval between observations to exploit the benefits obtainable from signal integration. Because the presence of even a very few ice crystals or water droplets can result in erroneous determination of molecular backscatter, complications exist in the selection of optimum beam cross sections. Further studies of the short-term density variations of the upper atmosphere are needed to assess how well widely spaced satellite lidar observations can represent general conditions.

In an effort to obtain a vertical atmospheric gas-density profile by selective attenuation, experiments are currently in progress to determine whether lidar systems can be made to function reliably at two wavelengths, one at some atmospheric-absorption band or line such as water vapor and the other at a nearby absorption-free wavelength of the spectrum. Theoretically it should be possible by this means to determine the density profile of the particular gas along the beam with such a system by comparison of the relative intensity of the returns at the two wavelengths until backscatter in the attenuated wavelength is reduced below measurable levels.

Brief experiments at SRI to temperature-tune an air-cooled ruby laser to the water-vapor absorption band at 6943.8 Å emphasized the difficulty of maintaining adequate temperature control (the absorption line is only a few angstroms wide). The difficulties introduced by the variation of the laser-rod temperature during its pumping cycle may be overcome by more efficient cooling.

With no observational experience yet available on which to base estimates of the accuracy with which density profiles could be measured or the depths to which the atmosphere could conceivably be probed from a satellite, the possibility of making this observation with a satellite lidar is indeterminable except on a theoretical and speculative basis at this time. It seems likely that the technical problems can and will be solved, perhaps quite soon, and this highly worthwhile experiment can be accorded a place in the list of potential lidar satellite observations.

The idea behind the observation of molecular structure by Raman-line observations is to exploit the Raman effect (shifts in molecular vibrational levels in gases and liquids when excited by electromagnetic radiation), which results in rotational emission lines at different wavelengths from the excitation wavelength. Given sufficient spectral frequency resolution and accuracy of determination of line intensity, information about the temperature and species content of a gas mixture can be obtained. Again, demonstrations that this is a practical observation in the free atmosphere using lasers have apparently not yet been made even at short range, so reliable data are lacking on which to base estimates of system power, sensitivity, and stability.

By radiational excitation of a gas at an appropriate wavelength its molecules may absorb and reradiate very strongly. Hence various gases in the atmosphere might be detected by a lidar radiating at appropriate wavelengths. The lidar must necessarily function in an absorption band or line of the spectrum, which of course attenuates the energy available for working at greater ranges than would be the case otherwise because much of the energy incident is absorbed and reemitted. This experiment, like those suggested above, is theoretically possible; however, no actual observational evidence that it can be accomplished at available laser wavelengths is known.

Because of the high degree of polarization of some types of laser-system beams, some investigators have wondered about the possibility of distinguishing between the backscattering produced by spherical cloud droplets of middle and lower cloud types and the asymmetrical, sometimes specularly reflecting ice crystals of cirrus clouds, since polarization techniques in microwave radar can eliminate the return from spherical raindrops. One would encounter some difficulty in making such measurements because of the depolarization of the beam along the path to and from the target.

While advances in electro-optical technology may shift the experiments suggested above from the "theoretically possible" to the "actually possible" and even "highly desirable" categories, it seems more worthwhile to concentrate our attention on the possibilities which have—at least at present—a more reasonable expectation of fulfillment in the foreseeable future. Considering those observations which appear to be possible given a lidar satellite with the capability of detecting Rayleigh backscatter at 10 km above the surface, we will now consider the extent to which this narrower class of experiments will be of interest to meteorologists, physicists, and operational users.

Putting a lidar aboard an experimental meteorological satellite would be worthwhile and interesting from a research or experimental point of view, just as was the first series of TIROS satellites. Prior to the advent of these weather satellites even the best high-altitude aerial and rocket photographs of cloud cover hardly hinted at the wealth of information which is now routinely obtained on a global basis from the satellites. Similarly, it is reasonable to expect that since almost nothing is known about the global occurrence and distribution of thin high-level cloud layers, a great deal of entirely new and fresh information about these difficult-to-observe clouds will become available with the advent of the lidar satellite. To the physical meteorologist the data should be of considerable interest because of the influence of these cloud systems upon incoming and outgoing radiation. The cloud physicist should gain new insight into the formation and dissipation of high cloud systems, and possible cosmic influences upon them, and the nature and extent of their influence upon precipitation formation at lower levels. The synoptic meteorologist will examine the data to learn more about the source and sink regions of high-level clouds and their relationship to upper-level air movement.

It is a virtual certainty that, because of the limited number of lidar and searchlight observations of the upper atmosphere, new and possibly significant optical-wavelength scattering layers may be found in the stratosphere and mesophere. *Gegenschein* and zodiacal light both provide evidence of substantial particulate concentrations at very high levels, and it is possible that other concentrations exist in such low density or in aposition with respect to the sun that they have heretofore completely eluded detection. While on this subject, we might note that the enormous sensitivity of lidar to particulate scattering offers opportunities for experiments involving the deliberate injection of highly reflective particles into the upper atmosphere for the observation of their subsequent motion. Ground-based lidar can readily detect very thin contrails of jet aircraft at a range of several miles.

To summarize, experimental or developmental lidar satellites could reasonably be expected to provide new data for study of the interrelationships between the mesosphere and stratosphere and possibly even the exosphere.

The operational value of lidar satellite observations will necessarily depend greatly upon the results of research using data obtained by developmental systems, that is, the correlations found between lidar satellite observations and important weather conditions which precede, attend, or follow them. Because of the numerous uncertainties involved, any suggested operational applications are speculative, and those set forth below are offered with this understanding clearly in view:

(1) jet-stream location by distinctive cirrus-cloud distribution,
(2) minimum and maximum temperature and frost forecasting as influenced by subvisible middle and high cloud layers,

(3) cloud-top determination and vertical structure of upper cloud systems for various aeronautical operations,

(4) height and structure of tropopause for meteorological analysis,

(5) horizontal and vertical water-vapor distribution (if such laser observations become possible),

(6) differentiation between overcast and clear snow-covered areas, and

(7) temperature and density determination from molecular scattering in the upper atmosphere.

Of course many of the above operational uses would require consideration of additional observational information such as TIROS-type cloud observations or Nimbus-type HRIR observations.

Users other than meteorologists may find operational applications for real-time lidar satellite observations. Photo and visual reconnaissance activities may be assisted in planning missions by more precise knowledge of the turbidity of the lower atmosphere or precise information on the height and thickness of cloud layers known to be present from TIROS cloud observations. By analysis of the polarization and intensity of the specular return from the sea surface, one may obtain some information concerning waves and the low-level wind velocity creating them. We will next investigate the rationale behind selecting cirrus clouds as the major lidar application and then consider briefly the meteorological significance of the cirrus formations.

During the initial phase of the SRI study a number of meteorologists both outside and within the Institute and with interests in both research and operations were polled for ideas on what a lidar satellite could and should do to justify its existence. Although the responses were of course quite varied and provided the basis for many of the applications listed above, the application which appeared on virtually every list was the mapping of cirrus cloud.

From an equipment point of view, the use of cirrus as the target for an initial satellite lidar effort offers several obvious advantages. It is the first visible feature of undisputed meteorological significance to be encountered when looking down from satellite altitudes; it occurs in systems of large (often global) extent; it is difficult to observe by any presently known technique since low cloud frequently masks it from view either from the ground or from satellite television. Finally, the scattering cross sections of typical cirrus clouds, while small compared with those of other clouds or even the normal aerosol content of clear air at sea level, are large compared with those of molecular constituents at any altitude.

In short, monitoring of cirrus is about the easiest job that can be envisioned for a satellite lidar. Since even this is a marginal proposition with present technology, until it is shown that cirrus can be successfully monitored, there seems to be little point to considering more elaborate systems. Early in the project, therefore, we decided to devote a major effort to defining the meteorological significance of cirrus cloud and to accumulate the data on physical and optical characteristics and global distribution required to evaluate the technical feasibility of monitoring it from satellite altitude with lidar.

During the study several factors justified this decision. The need for more quantitative data on cirrus and its value both to operational meteorology and radiation studies became much more obvious, and the technical problems that must be met in order to achieve even this initial goal became more clear.

Cirrus clouds are generated by meteorological disturbances that range from small-scale thunderstorms to large-scale tropical and extratropical cyclones. With thunderstorms, the cirrus appears as broad streamers of cloud flowing with the prevailing wind at high altitudes. With cyclones the cirrus appears as a gigantic sheet of cloud (as much has 1200 × 1200 nmi in area) over and in advance of the system. The upper tropospheric wind field with these systems may transport such cirrus thousands of miles downwind from its parent system. There is evidence that jet streams associated with major circulation systems not only transport cirrus but in many instances also contribute to its generation. Here the typical pattern is one of a cirrus band or series of bands some 50 to 500 mi wide, located on the anticyclonic or warm side of the jet stream (in the Northern Hemisphere, the right side looking downwind) and extending some 1000 to 2000 nmi in length.

On the left of Figure 2 is a high-resolution infrared (HRIR) picture of Hurricane Dora taken at 1:27 A.M. (EDT) by Nimbus I. The eye of the hurricane is clearly visible (dark circle in white cloud mass). Obscured by the storm's clouds are Florida, Georgia, North and South Carolina coasts, and almost all of the Gulf of Mexico. The Chesapeake and Delaware bays are just above the storm (at the top right).

The right-hand side of Figure 2 was taken by the Nimbus I Automatic Transmission Picture (APT) direct-readout camera. The storm is over the Florida coast, and although it is not as prominent as in the (HRIR) picture, the eye of the hurricane is again visible as a darker spot at the center.

Figure 3 is a Nimbus II APT picture showing a large double storm front over the East Coast and Midwest. The picture taken from an altitude of 700 mi shows a strip of storm clouds extending from Lake Huron to the southwest. Lakes Michigan, Huron, and Ontario are visible at the top.

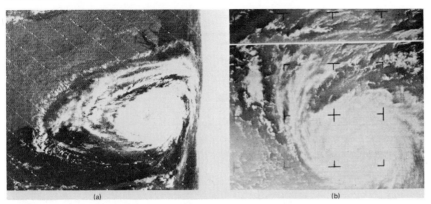

Figure 2 (a) High resolution infrared (HRIR) picture of Hurricane Dora taken at 1:30 A.M. by Nimbus I from an altitude of 245 mi. The Chesapeake Bay is at the upper right; (b) APT picture over Florida.

The two cirrus storm formations shown here cover millions of square miles and clearly have a direct connection with the present and future weather patterns for the areas beneath them.

Naturally, the hope arises that through a better identification and description, observations of cirrus can provide useful information on the nature and location of its parent system or on the nature of the upper-level wind field in which it becomes imbedded. If it can, then a device capable of providing detailed cirrus observations will become an important tool in analysis and prediction.

Of particular value would be knowledge of the intensity and structure of tropical or extratropical cyclones and future states of weather. Also, a description of the presence and characteristics of cirriform clouds would be valuable to radiation studies and to aviation, since cirriform clouds can be troublesome to such activities as jet aircraft refueling or rendezvous, celestial navigation, optical tracking, or photo reconnaissance.

To evaluate the range-finding capability of the lidar satellite in the detection of cirrus clouds, it is of fundamental interest to examine (on the basis of presently available data): (1) where and when cirrus clouds most frequently occur, (2) the associated configurations and dimensions of cirrus as seen from satellite altitude, and (3) what scanning patterns and spatial "sounding" coverage will be desired by an analyst and can be obtained from a satellite.

Complete data on the foregoing points have not yet been accumulated on a global or hemispheric basis for the following reasons: (1) Cirrus often cannot be observed from the ground when it is believed to be most abundant (during bad weather conditions). (2) Aircraft observations are

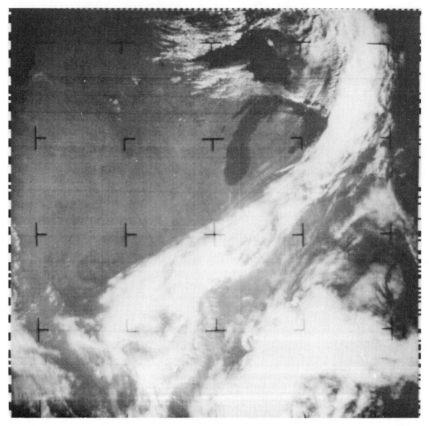

Figure 3   Nimbus II automatic picture transmission (APT) photo from an altitude of 700 mi. The picture covers an area from Florida to Nova Scotia and from the Atlantic to the Great Lakes.

available for only some specific situations and localities and therefore should not be generalized. (3) Cirrus, as it can be observed with presently available TIROS and Nimbus data, is difficult to map on a global or hemispheric scale because of inadequate satellite data coverage in space and time, not to mention interpretational problems.

Assuming that detection and mapping of cirrus clouds is desirable, a question then follows about the form and density of the observation. For the present, neglecting the considerable technical difficulties, we will assume that for the lidar (as for any active system) the cost and complexity of the required equipment will increase with the number of soundings required and that from satellite altitudes the absolute magnitude of the problem is such that continuous coverage with data density comparable to television or HRIR is out of the question. Thus, it is important at this point to examine from a meteorological viewpoint what is the minimum sampling density which could be useful.

In very basic terms, the operational meteorologist is interested first in the simple presence of cirrus cloud, and second in how much is present in what general area. Then he may become interested in the small-scale conformation. The latter is of much concern to the cloud physicist.

If large quantities of cirrus observations by satellite-borne lidar are to be made to evaluate the significance of cirrus in analysis and forecasting of tropical and extratropical cyclones, the density of data coverage should preferably be such that cirrus features on the mesoscale of the thunderstorm cell (3 to 10 nmi) can be detected and identified. This should be especially applicable to tropical and subtropical areas where the convective cells and the associated cirrus-generated cloud forms are important in the intensification and development of tropical cyclones. On the other hand, as previously noted in Figure 2, cirriform coverage generated from a single tropical or extratropical cyclone of interest may extend over an area in excess of 1200×1200 nmi; with respect to jet streams, the cirrus often appears in bands of extended length (2000 mi or more). Coordinated data coverage should be extensive enough that complete systems of this size can be examined.

Although one data point every 3 to 10 nmi (in both the north-south and east-west directions) can be accepted as an upper limit of data-point density required in cyclone analysis, the lower limit should probably not be less than one data point per 50 to 60 nmi^2. The minimum scale features that can be analyzed from a data coverage of one point per 50 to 60 nmi^2 are compatible with the scale of the convergence zones and frontal zones of a cyclone system. These zones constitute an essential part of the tropospheric models that are used to describe the weather and that are currently being analyzed with the standard meteorological observations (for example, frontal cyclone model).

The optimum method will be dependent not only upon the purpose for which the data will be used and upon an agreement on what is "significant" in cirriform cloudiness, but on engineering factors as well. However, from the standpoint of meteorology, it seems likely that the operational meteorologist will use the height data to construct charts of isopleths of cirrus altitude and thickness to be used in conjunction with the concurrent satellite photograph. He naturally will want, if possible, to know something about all the cirrus he can see in the photograph, and hopefully something about subvisible cirrus. Later, as information and knowledge accumulates, he may be in a better position to utilize fine altitude resolution from a smaller number of samples.

If initial power restrictions do not permit wide-area scanning the research meteorologist will certainly welcome lidar data taken simply as a series of soundings made along the satellite subtrack. A downward-pointing lidar capable of firing approximately once per second could generate a cloud altitude and thickness profile based on sample points taken every 3 to 3.5 nmi. This density of data acquisition is greater than that ordinarily achieved in atmospheric sampling and, if used in conjunction with a concurrent television or HRIR picture, could be of great value in evaluating the potential of the lidar and in planning future programs.

The minimum data density which would appear to be worth considering seriously for incorporation into an unmanned satellite even for research purposes would result from a vertical sounding about every 50 to 60 nmi along the orbital track, that is, approximately one every 18 sec. If the lifetime of the laser becomes an important consideration, it would be satisfactory to operate such a simple system by command from the ground only during times when interesting cloud situations prevail. However, given a sufficiently large power supply on the lidar satellite, the present ruby laser system should be capable of radiating an energy of 1 J every 1.8 sec, which is a factor of 10 better than the minimum frequency of one sounding every 18 sec.

Since the lidar would normally be used in close conjunction with other remote sensing devices contained in the same orbiting vehicle, it would seem presumptuous to assume that the orbital parameters would be optimized especially for the lidar. Fortunately most of the design criteria pertaining to orbit selection for television and high-resolution infrared systems are also valid for the lidar.

The ideal orbit for routine meteorological observation appears to be circular, near-polar, and retrograde (sun-synchronous). When provided, in addition, with an earth-oriented platform, a satellite in such an orbit provides global coverage with a high degree of day-to-day uniformity in viewing geometry and lighting conditions. If one assumes that the east-west field of view of a polar-orbiting observational satellite will be designed so that on consecutive orbits the mapped swaths will be contiguous at the equator, then it turns out that the amount of earth area passing through the field of view of the orbiting sensor per unit time is very nearly independent of satellite height. This is because the width of the swath required for contiguous coverage is directly proportional to the orbital period, whereas the distance covered in the orthogonal or north-south direction is inversely proportional to the orbital period.

Aside from the consideration of generally lower launching cost for lower orbits, the only argument in favor of a low altitude for a lidar satellite appears to be the $1/R^2$ factor in the lidar equation. Since transmitter power will always be at a premium in an active system, the orbit obviously should be no higher than is absolutely necessary to fulfill the other system requirements. However, if the altitude of the lidar satellite is too low, wide-angle lenses would be required for cameras on board the satellite, and because of the greatly extended path of the beam through cloud layers, the lidar resolution at the edge of the swath becomes too low. This effect for very low angles is shown in Figure 4.

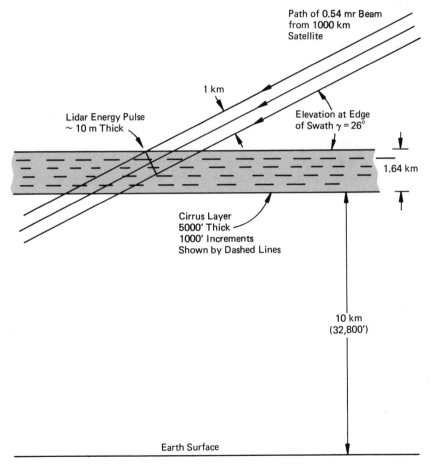

Figure 4 Beam geometry at edge of swath showing decreased Lidar resolution.

It appears that a 1000-km orbit with a constant earth-pointing axis on the satellite would be most favorable for the first experimental lidar mission. Later lidar satellites with more power available for the laser pulses may work more efficiently at higher altitudes.

Another important aspect of a lidar system is the earth terminal processing and user display of the lidar data. The lidar product at the telemetry receiving terminal could be a magnetic tape containing approximately 7.6×10^6 bits per orbit. Assuming 555 bytes per inch packing density and six data bits per byte, the data from each orbit would require 190 ft of tape; a typical 3600-ft roll of tape would accommodate lidar data covering about 33 hours of operation.

Depending upon the urgency and the available transmission facilities, the data could then be mailed or transmitted via wire or radio to a central computing facility. The computer would then be used as required to transform the raw data into plots of backscatter coefficient β'_{180} as a function of altitude or into descriptions in terms of inferred physical parameters such as particle size and number density.

The fact that each lidar sounding results in a complete vertical profile rather than a single value complicates the problem of data display. Considering how little is known about how the lidar soundings would be used, it appears unrealistic to attempt to describe an optimum display at this time. However, it is very important that we be able to describe one or more suitable displays, in order to permit meteorologists to envision what the lidar might do for them. Ultimately, of course, we look forward to the time when routine forecasting will be done entirely by the computer, and any output maps would be machine-prepared after proper weighting of data from many input sources, possibly including a satellite-borne lidar. In this idyllic extreme, the lidar input would have become so diluted that its individual contribution could seldom, if ever, be recognized by inspection of the output display.

At the other extreme, a cloud physicist performing research on a particular problem in cloud genesis might well be most satisfied with a tabular printout of the raw signal-return levels as a function of time with some set of specified limits on altitude and geographic location. Between these two extremes lie a variety of much more difficult display situations which must be designed to make lidar information available to a human analyst in predigested, on-call form. Like any good servant, the data should be available when needed but unobtrusive when it is not. One approach would be to print in the margin of a television or HRIR print small graphs showing β'_{180} as a function of altitude for critical points in the picture—in much the same way that vertical temperature profiles taken by ground stations located in the picture field are sometimes keyed into the TIROS television pictures.

A 12-digit all-numeric format which could be implemented by computer is shown in Table 1. Altitude and thickness are given in feet rather than meters to reduce the number of digits required for description in useful increments and over a useful total range.

TABLE 1

A 12-DIGIT ALL-NUMERIC FORMAT FOR TRANSMITTING LIDAR INFORMATION WHICH COULD BE IMPLEMENTED BY COMPUTER

Cloud Density on a Scale of 5	Cloud Thickness in Thousands of Feet	Height of Cloud Tops in Thousands of Feet
5 very dense		
4 dense	3 5 32	high (above 20,000 ft)
3 moderate	5 2 12	medium (10,000 to 20,000 ft)
2 thin	X X XX	low (below 10,000 ft)
1 very thin		

X's indicate data
 unknown because of
 absorptive loss in
 higher layers

In order to give the reader some impression of the data density involved even in such a highly reduced format, Figure 5 shows a TIROS photograph with an overlay gridded in 100-km^2 to correspond to a reasonable size of lidar reporting areas. Near the center of each square is a number group following the format shown above and representative of the reduced data which might have been obtained from a lidar sounding at a location indicated by the black dot immediately to the left of the number group. The photograph of Figure 5, taken over the west coast of Africa, near Cape Blanc, was chosen as an example for several reasons. It was made when the TIROS camera happened to be pointing straight down, which makes it representative of photographs taken from a stabilized platform such as that proposed for the lidar satellite. The optical contrast between the ocean and the desert is great enough to yield a striking similarity to a textbook map of the area. This contrast also demonstrates how much easier it is to see cirrus patterns against a dark background. Once the analyst's eye has been drawn to the area by the pattern of clouds over ocean it is not too difficult to follow the pattern on over the desert. However, had the dark patch not been there, the cirrus layer over the desert

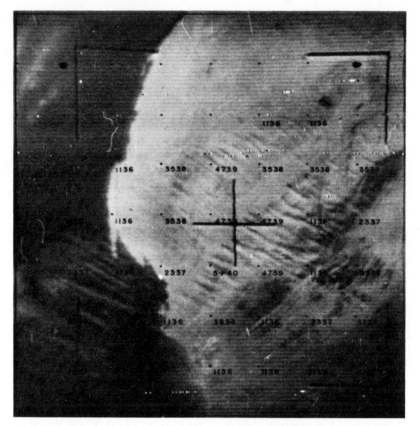

Figure 5   Simulated Lidar data superimposed on TIROS cloud photograph.

might easily have been missed by observation of the television picture alone. The range soundings provided by the (hypothetical) associated lidar confirm that the ripples in the desert picture are indeed caused by the same cloud system that is moving in from the sea. Bear in mind that the horizontal coverage is only 900 km, about one-third of the swath width proposed for a full-coverage, 1000-km, polar-orbiting satellite. (The + symbol in one of the thickness data positions on Figure 5 indicates that the cloud was 10,000 ft or more thick at that point.)

In conclusion, a pulsed ruby lidar capable of radiating 1 J per pulse and having an effective receiving aperture of at least 1 m^2 would be required to yield meteorologically useful soundings of simple backscatter as a function of range from a 1000-km orbit. The minimum pulse rate which should be considered for a full-coverage (scanning) system is approximately two per second and for a bare-minimum, downward-looking system approximately one every 18 sec. The equipment requirements for all other applications considered—including principally the measurement of gaseous temperature, density, or composition by spectroscopic means— call for orders of magnitude more power and/or receiving area, and for much more stability and calibration. Although there is no denying that the desire for remote measurement of temperature and density remains at or near the top of the meteorological "wish" list, competitive employment of satellite lidar for such applications must be considered only as an extremely remote possibility, at least until the practicability of the simpler cloud-monitoring system has been demonstrated.

Current lidar technology appears capable of equipping a meteorological satellite with the ability to make at least a few soundings per orbit of cloud densities and altitudes. Such soundings are useful for both operations and research and are very difficult to obtain by any other means. The prospects are good that improved laser efficiencies, output powers, and reliabilities will permit better coverage and more sophisticated measurements to be made in the future. It therefore seems desirable to begin serious planning toward a meteorological satellite lidar, having a minimum capability of ranging on cirrus clouds.

The project seems well adapted to a stepwise development sequence involving, first, the obtaining of additional design data via measurements from ground-based lidars; second, selective sounding of high cloud from above by a relatively simple lidar carried in a high-altitude aircraft; third, operation of basically similar equipment from a manned satellite; and fourth, incorporation of scanning and nonscanning systems into unmanned satellites.

JAMES O'BRIEN

James O'Brien received a B.S. degree in physics from Stanford in 1965 and, until he was killed in an automobile accident in March 1968, was working there for the Ph.D. degree in applied physics. He was a valuable member of the SPINMAP project and played an active role in the design of its remote sensing system. He worked for three years in laser research in the Hansen Laboratories of Physics at Stanford.

Abstract—*This article, entitled "Remote Sensors," concludes Part 7 and the book. It has been drawn from the Stanford SPINMAP report, from the ESSA publication "Man's Geophysical Environment" (1967) and has been written by Mr. O'Brien from Stanford.*

The balloon systems described earlier in this part by Mr. Ruttenberg and Dr. Morel are the first step toward the final global weather network. Ultimately, satellite-borne sensors will be able to measure temperature, pressure, water-vapor content, and air speeds remotely and there will be no need for the complex network of free-floating balloons now in preparation. But most of these sensors, like the lidar described by Dr. Ligda above, are in their infancy. They are design concepts or at best in the early stages of testing.

The numerical models need profiles of temperature, density, and so forth up through the atmosphere. Hence the sensors must be capable, not only of measuring the temperature, but of measuring the temperature at a predetermined altitude or pressure. It is the purpose of this article to discuss the theory behind such profile measurements and to outline some of the current proposals for the systems required.

Remote Sensors

IN the previous articles, meteorological measurements using sensors directly in contact with the atmosphere were described. These measurements were accomplished from ground stations, ships, rawinsondes, dropsondes, and an extensive constant-level balloon and buoy system. The disadvantages of these direct-sensing systems are numerous: cost, hazard to aircraft, poor worldwide coverage, and point- rather than volume-averaged data. A more attractive method of sensing meteorological parameters would be a completely remote satellite system that could yield all the upper-air data needed by both the numerical model and the synoptic meteorologist directly. A properly situated network of satellites could provide comprehensive coverage of the entire earth with whatever frequency the meteorologist requires, at transmitting data to the earthbound user without the delays which often degrade data from direct-sensing systems. Because of its position outside the atmosphere, a satellite sensor can be designed to provide spatially integrated data representative of the numerical model's earth-grid, as well as high-resolution information corresponding to that presently obtained from direct sensors in the atmsophere.

There are a great many quantities which might usefully be measured by satellite-borne remote sensors. Chief among these are atmospheric profiles of temperature, pressure, and water vapor, a determination of the earth's heat budget, the surface temperature of land and sea, sea-state determination, measurement of the snow pack, and continuous mapping of the distribution of ice in ocean and inland waterways. And as our understanding of the various meteorological phenomena becomes more complete, data from these remote sensors may yield invaluable information about the distribution of natural resources buried beneath the earth's surface. The potential of remote sensors as monitors of the earth and its atmosphere is far greater than that being explored in satellite systems of today. This article must be limited to only an introduction to the theory and application of remote sensing of meteorological data and an outline of its exciting future to the geologist, meteorologist, atmospheric physicist, and oceanographer.

THEORY OF REMOTE SENSING

A number of relatively simple physical processes determine what radiation will be seen by a satellite-borne sensor pointed at the earth. Without doubt the most familiar sensors of this outgoing radiation are the eyes and cameras of the astronaut, and the television "eyes" of the operational weather satellites. Sunlight passing through the earth's atmosphere is reflected from the ground, sea, and clouds, bouncing out again through the atmosphere to expose photographic film in the satellite camera, yielding a picture of the surface below. But the earth's atmosphere is not transparent to all electromagnetic radiation; wavelengths shorter than ultraviolet are either absorbed or scattered to such a degree that the earth's atmosphere is almost opaque to them. Transparency of the atmosphere increases with increasing wavelength as one moves into the large atmospheric "window" which we call the visible spectrum, extending from violet at 0.4 μ to the red and nearest infrared at about 0.7 μ. Numerous other spectral regions of transparency, or "windows," exist in the regions of the infrared and microwave, however, making such radiation also a potential tool for atmospheric sounding. We see then that, because our eyes are sensitive to radiation in the "visible" window only, we are led to believe that atmospheric transparency is the rule; rather, it is the exception.

Yet another kind of photograph is taken by the Nimbus High Resolution Infrared Radiometer (HRIR). Looking at the earth and clouds at night, when no reflected sunlight is present (or at sufficiently long wavelengths that the amount of reflected sunlight is small), these HRIR photos record the blackbody self-emission (thermal radiation) from the clouds, related through Planck's radiation law to the temperature of the emitting surface. *Planck's radiation law* is

$$E_f = \frac{2hf^3}{c^2(e^{hf/kT} - 1)}$$

where

E_f = monochromatic emissive power (W/m²/Hz)
h = Planck's constant
k = Boltzman's constant
c = velocity of light
f = frequency (Hz)
T = absolute temperature (°K)

In this manner remote sensors, looking through atmospheric windows in the infrared or microwave regions, can determine land, sea, and cloud-top temperatures.

These sensors while extremely valuable to any remote-sensing system, have been discussed in earlier articles. In this article we will explore the more advanced sensors using radiation of longer wavelength.

The infrared region comprises radiation of longer wavelength than the visible, and overlaps the microwave spectrum at wavelengths of about 1 mm. Yet another mechanism affecting the interaction of radiation and the atmosphere concerns the opacity or partial opacity of the atmosphere to infrared and microwave radiation; portions of the spectrum useful for passive sounding of the atmosphere are those in which the various gasses comprising the atmosphere absorb and re-emit radiation through molecular excitation. The various gasses such as CO_2, O_2, H_2O, CO, N_2O, O_3, and CH_4 absorb radiative energy passing through them at frequencies (or, when the individual spectral lines are not resolved, in frequency "bands") corresponding to the energy necessary to excite vibrations and rotations of their molecular structures. These bands usually overlap those of one or more of the other constituent gasses, the regions falling between absorption bands forming the atmospheric "windows." Within the individual narrow-frequency "lines," the molecular absorption frequency is dependent upon the temperature and density of the gas molecules. So in principle one should be able to determine the temperature of a gas, or an atmospheric constituent, by observing the frequency dependence of the radiation flux in one of its absorption bands.

Although the actual determination of the temperature profile in the atmosphere below a satellite is quite complicated, there seems to be little doubt that these difficulties can eventually be overcome, and that the atmospheric temperature profile can then be mapped from an orbiting spacecraft.

Several constraints determine the frequencies at which a meaningful temperature profile can be determined. First, one must observe the molecules of a gas in a band where other molecular mechanisms are not active (the problem is sufficiently complex with one gas, and considerably more so for several gasses). Also, the gas should be one evenly distributed in the atmosphere; water vapor, for example, would not be suitable because of its irregular distribution. If we are to obtain complete profiles, the radiation used should be capable of partial penetration to the earth's surface.

Both CO_2 and O_2 are found to be quite uniformly distributed through the atmosphere, and both have absorption spectra with active regions uncluttered by the effects of other gasses; CO_2 has a particularly good active band in the infrared at 15 μ, as oxygen does in the microwave at a wavelength of 5 mm (corresponding to a frequency of about 60 GHz; microwave radiation is most conveniently described in terms of frequency, rather than wavelength). Temperature-profile measurements could be made using a passive remote sensor in either the infrared or microwave regions, although several factors discussed later favor using the microwave oxygen lines.

Whatever the frequency choice, the theory of profiling atmospheric temperature from a set of measurements of energy flux at the top of the atmosphere is the same. The amount of flux reaching the top of the atmosphere at a particular frequency is given by:

$$N = \int_{\tau_1}^{\tau_2} B(T) \, d\tau + \tau_1 B(T_s)\epsilon_s$$

The first term is the contribution due to the atmosphere, and the second term is that due to the surface emission. τ is transmissivity of the gas, $B(T)$ the blackbody radiance of the gas at a temperature T, ϵ is the emissivity (the subscript s indicating surface), τ, is the total transmissivity of the atmosphere, and N is the observed flux or radiance. Now if we divide the atmosphere into layers, to calculate the flux it is necessary to know the emissivity of the surface and of each layer of the atmosphere. It can be shown that, having sufficiently many measurements of N at different frequencies and being able to evaluate the effect of the second term due to surface emission, one can invert the radiative transfer equation to obtain a temperature assignment to each layer. In this manner one can obtain a profile relating atmospheric temperature to pressure altitude.

A more rigorous analysis of the theory of inversion of the radiative transfer equation must take into account scattering due to water droplets in the atmosphere, collisional broadening of the spectral lines, and the complexity of a mathematical solution of the resulting nonlinear integral equations; numerous papers are appearing in journals, and will be left for the most determined reader to explore. It is sufficient at this point to say that recovery of a unique solution requires at least n independent samplings to profile an atmosphere represented by a model having n levels or layers.

To determine a temperature profile in this manner, we require knowledge of the distribution of the gas (homogeneous, for O_2 and CO_2), and the total flux from the top of the atmosphere at several frequencies to arrive at a unique determination of the temperature of the atmosphere. Assuming that the temperature profile has been so obtained for an atmosphere, one could again measure the outgoing flux at a frequency corresponding to *any* atmospheric constituent, and, now knowing the temperature distribution, deduce the distribution of that gas in the atmosphere. This is precisely what we wish to do in the case of water vapor. Conducting measurements on the diatomic oxygen lines and water-vapor lines would yield profiles of both atmospheric temperature and water vapor versus pressure altitude.

The various remote-sensing techniques discussed so far have been passive; that is, they "see" and measure radiaton coming to them from the sun through reflection, from blackbody emission, and from molecular absorption and emission. Another tool for remotely sounding the atmosphere involves active devices which transmit signals to probe to the earth's surface. These devices will be considered in a later discussion of radio occultation.

INSTRUMENT DESIGN

As mentioned earlier, the device to be used as a remote sensor will be determined by the phenomenon to be observed and by the electromagnetic radiation whose wavelength (frequency) is most suitable for that observation. Because infrared radiation has been used as a scientific tool for a longer period than the microwave has, to date most sensors flown have been devices measuring visible and infrared radiation effects. Useful information on the albedo of the sea, ice, land, vegetation, and clouds, as well as pictures, can be gathered using near-infrared radiometers at "window" frequencies. An atmospheric window between 10 and 11 μ is particularly well suited for measurement of ground and cloud temperatures. But semiopaque cloud cover degrades data, and opaque clouds prohibit infrared study of the atmosphere and surface below them. The presence of cirrus clouds can seriously degrade measurements of thermal flux; studies indicate that, at infrared wavelengths, cirrus clouds make profiling below their level difficult. And photographic studies show that cirrus clouds are present 40 to 50 percent of the time, and at altitudes up to 50,000 ft. The presence of clouds prohibits complete atmospheric profiling to the ground through the spectral region from 4 to 1000 μ. This inability to penetrate cloud cover means other sensors would have to fill in for complete global atmospheric profile measurements.

Microwave radiometric techniques are not limited to clear-sky conditions, however. Studies by D. Diermendjian for the RAND Corporation indicate that millimeter microwaves penetrate liquid water and ice crystal clouds, and, to some extent, rain. The more significant absorption and scattering effects in raining clouds could be used to monitor heavy storm activity.

Other factors must be considered. The radiation in the microwave region varies directly as the temperature of the gas, while in the infrared region radiation varies as the fourth power of the gas temperature. This variation results in a potentially better vertical resolution for the infrared than for the microwave. It is also more difficult for microwave systems to separate ground emissions from emissions in the lowest layers: The latter difficulty is partially offset, however, by the ability of microwave receivers to measure emission between absorption bands, a capability not available in infrared detectors. Penetration of cloud cover is still the overwhelming argument for the development of microwave systems to see where infrared systems are blind. Microwave will still be effected by some absorption of strong rain clouds; however, it seems possible to correct for this by prediction of rain droplets from other parameters or by measurement of rain absorption on the signals of the occultation system described below.

About thirty individual oxygen resonances can be observed in the microwave interval between 50 and 66 GHz. As noted earlier, measurements taken in this frequency region will be composed of contributions from the oxygen resonances, emission from the earth's surface, and perhaps a minor contribution from sulfur dioxide or ozone. One separates these by making additional simultaneous measurements at frequencies isolating particular components; observations at "window" frequencies of 35 and 95 GHz, on either side of the oxygen resonances, would allow one to isolate the temperature-bearing oxygen data in radiometric measurements near 60 GHz. In addition, one might use an H_2O resonance at 22.2 GHz to obtain the water-vapor distributon, as suggested earlier.

In addition to providing calibration for the temperature and water vapor profiles, data gathered from the 35-GHz region would yield information on the distribution of sea ice, thickness of level ice over water, thermometric temperature of the sea, distribution of condensed water vapor over water, and could provide a measure of the sea state and the sea swell factor. Aircraft and satellite tests of microwave radiometers in the 22-, 35-, and 60-GHz regions should be made at an early date to develop these important remote sensing techniques.

Passive microwave measurements could provide profiles of temperature, and water vapor versus pressure altitude, as described in preceding sections. A radio occultation system could provide a mass-density profile as a function of height. Combining the two sets of data would give, for the first time, operational profiles of temperature, pressure, and water vapor height.

The satellite occultation experiment is based on the principle that the refractive index of the atmosphere is directly proportional to its mass density. By measuring the effect of a changing refractive index on an electromagnetic wave passing through the atmosphere, it is possible to determine the corresponding change in mass density.

Two methods of making this measurement have been investigated. The first, stellar occultation, involves making satellite measurements of the refraction of a ray of light as the star is occulted by the atmosphere. The refraction angle is a maximum of about 22 mrad when the low point of the ray is at sea level. If this angle can be measured to ± 0.01 mrad, density profiles with accuracies comparable to that of rawinsondes are possible. There are two approaches to this idea. The first involves using two star trackers on a low-level earth-oriented satellite. Each star tracker locks onto a star and the change in relative angle between them is measured as one of the stars is occulted. This method has disadvantages mainly because of the high stability required on the satellite. The second approach utilizes a low-level rotating satellite with its spin axis perpendicular to the orbit plane. The sensor for this satellite is a telescope mounted radially followed by a set of thin slits and a photomultiplier tube. As a star image is swept past the slits, light pulses are sent into the phototube, which produces electrical output pulses. Any changes in the relative spacing of these pulses can be attributed to a star's being occulted by the earth's atmosphere. This method can achieve the necessary accuracy, but is of questionable value because the presence of clouds will normally limit the minimum altitude of the measurements to the 300-to-500-mb level.

A radio occultation system which propagates microwaves between two satellites so that the radio wave passes through the earth's atmosphere effectively overcomes this difficulty. The main advantage of using microwaves is that they can penetrate clouds, and it is theoretically possible to correct for any effect the cloud has on the transmission path. Also, extremely precise Doppler measurements on the relative velocities between the two satellites can be made by using developed deep-space tracking techniques.

The basic idea of the radio occultation satellite system is straightforward and is shown in Figure 1.

As the radio signal passes through the atmosphere, it is refracted, and this in turn gives rise to a change in the apparent motion between the

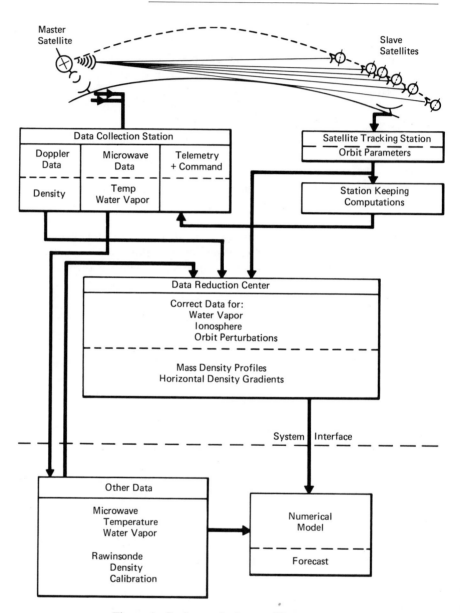

Figure 1   Basic occultation satellite system.

two satellites. If all other effects that produce a Doppler shift are accounted for (such as any real motion between the satellites due to perturbations in the orbit, or any other apparent motion caused by refraction due to the ionosphere and water vapor), the remaining changes can be attributed to refraction by the atmosphere which can in turn be related to change in the atmosphere mass density.

The radio occultation experiment can be implemented with three major subsystems: a platform, a radio transponder, and a data reducton center. The basic platform consists of one master satellite and six repeater slave satellites in the same highly circular polar orbit. The slave satellites are spaced so that radio waves propagating between them and the master satellite will cut the atmosphere at a number of different levels. Since intersatellite spacing is very critical, all satellites will require station-keeping ability. The radio transceiver and electronics are mounted on the satellite to measure and record Doppler shifts and send the data to the ground collection center. Since a Doppler shift can be caused by a number of other effects besides changes in mass density, provision must be made to remove these errors from the data at a ground data reduction center.

The atmospheric contribution to the measured Doppler shift is due to two effects. First, the ray is retarded by the atmosphere, and second, the ray is bent and therefore travels a longer path. Making the assumption that the atmospheric mass density is exponentially distributed with altitude allows the size of these two effects to be determined. Assume

$$\rho = \rho_s e^{-z/H} = 1.200 e^{-z/8.1}$$

where

z = altitude (km)
H = scale height (km)
ρ = density (kg/m³)
N = modulus of refraction

Since

$$N = 222\rho + [\text{water-vapor term}]$$

we have

$$N = N_s e^{-z/H} + [\text{water-vapor term}]$$
$$N = 270 e^{-z/8.1} + [\text{water-vapor term}]$$

With these assumptions, the calculations are relatively straightforward and are covered in detail in numerous sources where the ideal nature of the atmosphere is also considered. If

$\Delta R(h)$ = retardation effect as a function of altitude
$\Delta L(h)$ = bending effect as a function of altitude
$\Delta \epsilon(h)$ = bending angle as a function of altitude
$\dot{\delta}d$ = apparent Doppler shift
h = altitude of measurement
h_0 = altitude of straight-line path
d = deviation of true path from ideal

then

$$\dot{\delta}d = \frac{d}{dt}[\Delta L(h) + \Delta R(h)] = \frac{d}{dt}\left[Ra\frac{\epsilon^2(h)}{2} + \Delta R(h) \right] \qquad (1)$$

The result is

$$\epsilon(h) = 2 \times 10^{-6}N_s e^{-h/H} \cdot \frac{1}{H} \cdot \left(\frac{\pi RH}{2}\right)^{1/2} \approx 21.5e^{-h/8.1} \qquad (2)$$

$$\Delta R(h) = He(h) \approx 348e^{-2h/8.1} \qquad (3)$$

$$\Delta L(h) = \frac{Ra\epsilon^2}{2} = 810e^{-2h/8.1}$$

$$\int \dot{\delta} \, dt = 348e^{-h/8.1} + 810e^{-2h/8.1}$$

$\epsilon(h)$ is in milliradians, h, $\Delta R(h)$, $\Delta L(h)$, and $\int \dot{\delta} \, dt$ are in meters.

This is the total apparent distance change caused by the mass of the atmosphere. At sea level

$$\int \dot{\delta} \, dt = 1258 \text{ m}$$

and

$$e_s = 1.2 \text{ kg/m}^3$$

Therefore, a 1-m change in apparent distance is equivalent to a 1 g/m³ change in mass density. The proposed Doppler system operates at 5.25 GHz and can resolve changes in apparent distance of λ = 4.70 cm. Therefore, detecting a 1-m apparent distance change presents no difficulties. The diffculties in the experiment arise in eliminating other sources of Doppler shift and in determining the height of the measurement. Figure 2 shows the geometry of a typical occulation problem.

Other sources of Doppler shift which must be accounted for are due to oscillator stability characteristics, the ionosphere, water vapor, and relative satellite velocity due to orbital eccentricities, differences of period, and harmonics of the earth's geopotential. Taking the various error sources into account, it appears possible to measure mass density profiles to within 1 percent and to measure horizontal density gradient to within 0.2 percent.

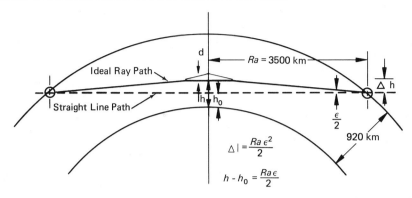

Figure 2   Doppler shift geometry.

It is essential to know the global distribution of atmospheric flow as an initial condition for numerical weather prediction. Even if it were possible in middle and high latitudes to do without this wind information by applying the geostrophic approximation of the massfield,* the tropical zones would still require the horizontal velocity as the primary observational parameter. Conventional methods to determine upper winds require in situ measurement by balloons rising or drifting in the atmospheric flow. To achieve this on a global scale with sufficient horizontal and vertical resolution requires a very large system indeed. The World Weather Watch is considering, as one alternative, a Global Horizontal Sounding Technique (GHOST) using balloons which float at constant density surfaces (as discussed earlier in this part by Dr. Hallgren).

For later phases of the World Weather Watch a remote-wind-sounding system would be more attractive. The proposal described here originated at NASA's George C. Marshall Space Flight Center (Dr. F. Krause). It is interesting enough to warrant a feasibility test program. If successful, it would be a simple and elegant way to determine worldwide atmospheric winds from space and would be of extraordinary significance for tropical meteorology.

The proposed test arrangement ("crossed-beam method") employs two orbiting remote scanning devices with triangulation between two lines of sight to select a desired altitude. This "crossing height" is defined by the minimum distance between the almost intersecting lines of sight (Figure 3). The detectors measure passive radiation in the atmosphere.

---

* The material in the paragraphs on wind measurements has been taken from the ESSA publication *Man's Geophysical Environment—Its Study from Space* 1967, with the kind permission of the editor, Dr. J. P. Kuettner.

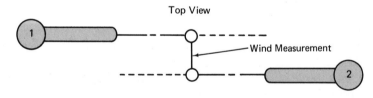

Figure 3 "Crossed beam" method for remote measurement of wind. At minimum distance of the two lines of sight, radiative fluctuations in the atmosphere have maximum cross correlations yielding average wind velocity at "crossing height."

Each orbiting detector monitors all light emission and extinction processes along its line of sight. As meteorological disturbances cause modulations of the detected radiative power, the information related to all meteorological fluctuations along the line of sight is obtained. Assuming that the fluctuations have maximum cross-correlation at the "crossing height" the local information at this height can be retrieved and the remaining signals in the line of sight ("flow noise") eliminated by a digital, real time cross correlation routine. Analytical studies and experimental wind-tunnel tests indicate that:

(1) Convection velocity, length scale, and other parameters may be resolved, normal to the plane of the beam, within limits that are given by statistical errors and the beam diameter (approximately 50 m).

(2) Passive radiation from meteorological modulations within the line of sight provides enough radiative power to give optical signals in excess of external and detector noise levels.

(3) The measured convection speeds are group velocities and will differ from mass average (wind) velocities only if positive and negative fluctuations of the optical signal are not equal. However, wind-tunnel experiments under extreme conditions (free shear layers) show that the difference should not exceed a few percent of the wind speed.

In case of mountain-induced stationary waves or vortices correct wind values can probably not be expected. The method would be most useful over vast ocean areas where aerological data are missing.

The concept applies equally to the visible, the infrared and the micro-wave region of the spectrum and might be used to measure flow velocities inside and outside clouds.

ADDITIONAL REMOTE MEASUREMENTS

The heat exchange between the earth and surrounding space is not uniform. More heat is absorbed at the equator than is emitted there, and the opposite picture is true at the poles. In order to maintain a heat balance, energy must travel from the region of the equator to the region of the poles. This transfer of energy is the driving force of the atmosphere. A thorough knowledge of the amount of heat radiating from all parts of the earth should therefore lead to a much better understanding of the movement of the atmosphere and the weather in general.

In order to learn how much energy is being absorbed at the equator and how much is being emitted at the poles, a rapid sampling of the energy flux to and from the earth must be taken. The means for this sampling is readily at hand with radiation-measuring satellites. Since the earth is heating and cooling rapidly as portions of it go through day and night, several satellites would be used, and data taken from the same location several times a day.

The simplest monitoring device could be two flat plates held parallel to the surface they are flying over and exposed to radiation coming from the grid spacing chosen. One of the plates would be painted black and therefore absorb all radiation (both reflected from sun and emitted from earth). The other plate would be white and absorb only radiation emitted from the earth (white reflects solar radiation which has been reflected from earth). Two identical instrument packages could be mounted on opposite ends of a gravity gradient boom. The gravity gradient system will keep one set of instruments in position over the surface of the earth and the other pointing at space. The package pointing at space would collect no useful data, but it is more economical to construct a system with two simple identical sets of instrumentation than to put a dead mass on one end of the boom and be sure that the instrumented package is pointed at earth and not at space.

These lightweight, simply constructed satellites could be placed in orbit in the near future from a polar-orbiting Apollo mission as a planned scientific experiment. Three might be placed in orbit during one mission and three more during a second mission which is 90° out of phase with the first. If a polar Apollo mission were not available, the satellites might be placed in orbit with a Scout rocket.

The ocean acts as a bank for the earth's heat budget. The atmosphere is essentially heated from the oceans as sea water supplies heat across the ocean-air interface. Since local heat sources over the sea supply energy to cyclonic disturbances at their origin, meteorologists would find a distribution map of those local ocean sources of considerable value. Measurement of the temperature gradient between the surface of the ocean which is radiating heat to the atmosphere, and levels at a 5-mm depth is a direct measurement of the heat flow from the sea. Such measurements are difficult, perhaps impossible, to make directly.

A passive microwave radiometer might be able to gather that data from a remote satellite position, however. An electromagnetic wave propagating in space and interesecting a medium with electrical properties different from vacuum will be, in part, reflected back from the interface. The remaining energy will be attenuated exponentially as it passes into the medium.

The extent of penetration is represented by a skin depth D for each wavelength, the distance in which the wave is attenuated by a factor of $(1/e)^2 = 0.135$. Blackbody principles of reciprocity also demand that radiation reemitted from the surface originates in the same distribution through the skin depth. We can therefore attribute the radiation emitted from the sea to a source whose depth is a derivable function. By properly selecting a number of microwave frequencies, a satellite radiometer should be able to monitor quite accurately the radiation density versus depth to a sea depth of 4 or 5 mm. The effects of variable sea state could be determined and removed by simultaneous microwave measurements at selected frequencies and polarizations.

The apparent temperature of the sea as seen by a satellite radiometer will be a function of the thermometric sea temperature, absorption due to the atmosphere, and a reflectivity term proportional to the product of apparent sky temperature and ocean reflectivity. Atmospheric absorption is small but for a small number of absorption bands; ocean thermometric temperature can be determined from vertical microwave soundings at two frequencies. The third term will vary with the tilt of the ocean surface and with the observation angle, since these factors determine the apparent reflected sky temperature, and the apparent radiometric temperature will be a function of multiple sky reflections at certain angles. There appears to be a good possibility of correlating the signatures of the various sea states with a microwave radiometer measurement of vertical incidence and a second at some grazing angle, say 50°. In addition to providing valuable information to the oceanographer and meteorologist, such data might allow ships to be routed around areas of ocean surface disturbance. Imaginative design of a radiometric sea state monitor might possibly yield information on surface winds at sea.

Because surface penetration of microwave radiation varies with wavelength, one could, in theory, monitor several frequencies to obtain a vertical temperature sounding of various portions of the earth's surface.

About 10 percent of the earth's surface (15,000,000 mi²) is covered with ice. Numerous water transportation lanes are troubled or closed at least a part of every year because of sea ice. The U.S.S.R. in particular attempts to use aircraft observations to chart ice coverage throughout the year, regularly duplicating flight paths. There is no doubt that an accurate knowledge of the ocean ice distribution would not only increase the amount of shipping in the course of a year, providing data for icebreakers, but also would decrease the danger to ships from ocean ice. One can envision a Baltic shipping season extended for a week or more, with a proportional increase in tonnage moved.

Present visual methods of observation are hampered by severe weather, fog, and short days, the very conditions characteristic of the area in which observations are needed. A satellite-borne microwave radiometer could operate in the heaviest of fogs and allow a survey of sea-ice distribution to a far more accurate scale. A truly accurate ice map of the Arctic could be made and kept up-to-date at a cost considerably below the economic return.

The principle on which such a device would operate is the following. The apparent surface temperature seen by a microwave radiometer is dependent on intermediate atmospheric attenuation, surface thermometric temperature, and surface reflective and emissive characteristics. The reflective and emissive characteristics of sea water and sea ice are so at variance that monitoring the apparent temperature differential ΔT seen by a satellite microwave radiometer would be a simple task. Microwave radiometers can easily monitor ΔT's of a few degrees Kelvin; and so favorable are the temperature differences between ice and sea water that one could easily construct a radiometric ice map for any desired coverage.

Table 1 illustrates a sampling of tabulated values for this apparent differential temperature between sea water and ice:

TABLE 1*
TOTAL VERTICAL ATTENUATION AND APPARENT SKY TEMPERATURES AT 35 GHz

	Total Vertical Attenuation	Apparent Sky Temperature	Apparent ΔT
Clear day 1 g/m³ water vapor	0.2 dB	17°	120°
Light fog 1000 ft visibility	0.4 dB	33°	106°
Heavy sea fog 100 ft visibility 2000 ft thick	0.6 dB	49°	100°
Snow storm 5 mm/hr Water and ice cloud	1.5 dB	108°	51°

* Courtesy of Space General Corporation

The receiver frequency suggested for an ice-mapping instrument is 35 GHz, falling near the atmospheric absorption minimum between the 22-GHz water vapor and the 60-GHz diatomic oxygen absorption peaks. This frequency is high enough to allow a choice of beamwidth sufficiently narrow for maximum ground resolution to be achieved with a practical antenna size. A sample calculation will reveal that the necessary detector sensitivity can be easily achieved.

Because the areas where ice cover is a problem are within about 30° latitude of the poles, an ice monitor in polar orbit (with its inherent redundancy over precisely that area) would be ideal. Furthermore, since the ice phenomenon is of relatively low frequency, one could tolerate a nonscanning antenna and still obtain sufficient ground resolution to update an arctic ice map at least once a week; one need only have successive orbital passes offset by the antenna beam coverage at its position of lowest latitude of observation. The necessary microwave equipment could be contained in a small and light enough package to allow its adoption as a passenger to many of the large number of polar orbiting host satellites which will doubtless be orbited in the future.

In addition to its ice mapping duties, this instrument could provide surface temperature data to whatever accuracy one wished to buy from the sensitivity-beamwidth-dwelltime tradeoff. A 35-GHz receiver might be used to calibrate a simultaneous temperature profile experiment, or provide data to determine global sea state. The potential of a microwave mapping device would be greatly clarified by inclusion of the 35-GHz experiment on an early Nimbus spacecraft.

Bibliography

In order to assist those interested in reading further on the subject, the following bibliography has been prepared. The references are divided into categories which correspond approximately to the sections of the book. The World Meteorological Organization (WMO) is one of the best sources on all aspects of weather prediction.

GENERAL

"Weather Satellite Systems," *Astronautics and Aerospace Engineering,* Vol. 1, No. 3, 1963.

Rapp, R. R., and R. E. Huschke. *Weather Information: Its Uses, Actual and Potential.* Santa Monica, Calif.: The Rand Corporation, 1964.

Sawyer, J. S. "Performance Requirements of Aerological Instruments," Technical Note No. 45, WMO No. 119, TP 54, 1962.

WMO Bulletin, published quarterly. Information about subscriptions and issues may be addressed to the Secretary-General, WMO, Case Postale No. 1, 1211 Geneva 20, Switzerland.

WMO. Commission for Aeronautical Meteorology: Abridged Final Report of the Third Session, WMO No. 157, RP 58, Geneva, Switzerland, 1964.

WMO. *Automatic Weather Stations.* Tech. Note No. 52, WMO No. 136, TP 62, 1963.

WMO. *The Present Status of Long-Range Forecasting in the World.* Technical Note No. 48, WMO No. 126, TP 56, 1962.

THE WORLD WEATHER WATCH

Nat'l. Acad. Sci. *The Feasibility of a Global Observational and Analysis Experiment,* Washington, D.C.: N.S.A. Publication 1290, 1966.

Nat'l. Acad. Sci. *The Global Atmospheric Programme (GARP),* Report of the Study Conference held at Stockholm, 28 June–11 July, 1967. Cosponsored by the WMO.

WMO *The Role of Meteorological Satellites in the World Weather Watch,* World Weather Watch, Planning Report No. 18, 1967.

WEATHER MODIFICATION

Nat'l. Acad. Sci. *Weather and Climate Modification Problems and Prospects,* Vol. 1. *Summary and Recommendations,* Vol. 2. Research and Development Pub. No. 1350. Washington, D.C.: N.S.A. 1966.

Zikeen, N. T., and G. A. Doumani. *Weather Modification in the Soviet Union, 1946–1966: A Selected Annotated Bibliography.* Washington, D.C.: 1967.

THE ECONOMICS OF WEATHER PREDICTION

Demsetz, Harold. *Economic Gains from Storm Warnings: Two Florida Case Studies.* Santa Monica, Calif.: The Rand Corporation (prepared for NASA, RM 3, 68), 1962.

Gringorton, I. I. "Forecasting by Statistical Inferences," *Journal of Meteorology,* Vol. 7, No. 6, December 1950, pp. 388–394.

Kolb, L. L., and R. R. Rapp. "The Utility of Weather Forecasts to the Rosin Industry," *Journal of Applied Meterology,* Vol. 1, No. 1, March 1962, pp. 8–12.

Lave, L. B. "The Value of Better Weather Information to the Rosin Industry," *Econometrica,* Vol. 31, No. 1-2, January–April 1963, pp. 151–164.

Senko, M. E. *Weather Satellite Study-A Special Report.* Washington, D.C.: U.S. Department of Commerce, Weather Bureau (unpublished).

SPINMAP—Final Report of Engineering 235, 1966, Stanford University, Stanford, California.

Thompson, J. C. "Economic Gains From Scientific Advances and Operational Improvements in Meteorological Prediction," *Journal of Applied Meteorology,* Vol. 1, No. 1, March 1962, pp. 13–17.

Thompson, J. C. *The Potential Economic and Associated Values of the World Weather Watch.* WWW Planning report No. 4, 1966.

WMO. *The Global Data-Processing System and Meteorological Service in Aviation,* WWW Planning Report No. 13, 1966.

WMO. *Meteorological Service for Aircraft Employed in Agriculture and Forestry,* Technical Note No. 32, WMO No. 96, TP 40, 1960.

WMO. *The Potential Contribution of the World Weather Watch to a Global Area Forecast System for Aviation Purposes,* WWW Planning Report No. 19, 1967.

WEATHER SATELLITES AND COMMUNICATIONS SYSTEM

Akima, Hiroshi. *Speeding up of HF Radiotransmissions and Broadcasts,* (World Weather Watch Study T. 21) U.S. Dept. of Commerce (ESSA), 1967.

Berry, F. A., P. T. Willis, and J. G. Gronin. *Operational Use of Weather*

Satellite Information, Final Report for Project FAMOS U.S. Fleet Weather Central, Suitland, Md., 1964.

U.S. Dept. of Commerce, ESSA, *APT Users Guide.* Washington, D.C.: Government Printing Office, 1965.

WMO *Planning of the Gobal Telecommunication System,* WWW Planning Report No. 16, 1966.

WMO. *Reduction and Use of Data Obtained by TIROS Meteorological Satellites,* Technical Note No. 49, WNO No. 131, TP 58, 1963.

Zver, G. A. *Plan for a Regional Telecommunication Network for Region II (ASIA),* WWW Planning Report No. 6, WMO, 1966.

NUMERICAL MODELS

WMO. *Numerical Methods of Weather Analysis and Forecasting,* WMO No. 118, TP 53, 1962.

The Physical Laws:

Bjerknes, V. "Das Problem der Wettervorhersage, betrachtet vom Standpunkte der Mechanik and der Physik," *Meteor, Zeitschrift,* Vol. 21, 1904, pp. 1–7. (English translation in ESSA, Weather Bureau, Western Region Technical Memorandum 9, 1966; The first explicit formulation of the weather forecasting problem as an initial value problem to be solved by graphical or numerical methods.)

Leith, C. E. "Numerical Hydrodynamics of the Atmospere," *American Mathematical Society Proceedings of the Symposium in Applied Mathematics,* Vol. 19, 1967.

Mintz, Y., A. Arakawa, and A. Katayama. "Numerical Simulation of the General Atmospheric Circulation and Global Climate," *Numerical Simulation of Weather and Climate* (Technical Report No. 3, Dept. of Meteorology). Los Angeles, Calif.: University of California, 1968. (An example of a numerical general circulation experiment by long-period integration of the primitive equations, with a coarse grid and simple parameterization of the heating and friction. It shows the degree to which the characteristic propertes of the winds, temperature, surface pressure, humidity, and precipitation can be simulated.)

Phillips, N. A. "Numerical Weather Prediction," in *Advances in Computers,* Vol. 1, F. Alt, Ed. New York: Academic Press, 1960, pp. 43–90. (A review of modern numerical weather prediction. It includes a discussion of prediction with the filtered (geostrophic) system of equations.)

Richardson, L. F. *Weather Prediction by Numerical Process.* Cambridge, Mass.: University Press, 1922. (Paperback reprint by Dover Publications, New York, 1965; reviewed by G. W. Platzman in *Bull. Amer. Meteor. Soc.* Vol. 48, 1967, pp. 514–550, an elaborately organized scheme for numerical weather prediction with the primitive equations, with many suggestions for parameterization of the sub-grid scale physical processes.)

Finite Difference Approximation:

Arakawa, A. "Computational Design for Long-Term Numerical Integrations of Fluid Motion: Two-Dimensional Incompressible Flow. Part I," *J. Comp. Physics,* Vol. 1, 1966, pp. 119–143.

Arakawa, A. "Mathematical and Computational Aspects of Numerical Simulation of the Atmosphere," in *Proceedings of the American Mathematical Society, Symposium on Numerical Solution of Field Problems in Continuum Physics,* 1968. (In preparation; discussion of finite difference schemes, with a stress on nonlinear computational stability.)

Marchuk, G. I. *Numerical Methods in Weather Prediction* (in Russian), 1967. (English translation in preparation for Holt, Rinehart and Winston, New York; a comprehensive discussion of the prediction problem. Emphasizes the "splitting method" for solving the equations.)

Mintz, Yale. *Very Long-term Global Integration of the Primitive Equations of Atmospheric Motion,* WMO No. 162, TP 79, 1965.

Sadourny, R., A. Arakawa, and Y. Mintz. "Integration of the Non-divergent Barotropic Vorticity Equation with an Icosahedral-Hexagonal Grid for the Sphere," *Monthly Weather Review,* 1968. (Example of a quasihomogeneous grid for the global domain.)

Specification of the Initial State:

Charney, J. G. "A Global Observation Experiment," in *Dynamics of Large-Scale Atmospheric Process, Proceedings of the International Symposium,* Moscow, June 23–30, 1965, pp. 21–35. See also *The Feasibility of a Global Observation and Analysis Experiment.* Washington, D.C.: National Academy of Sciences, National Research Council, 1966. (A discussion of the need for global weather observations. Discusses the limits of predictability.)

COSPAR Working Group VI, "Status Report on the Applications of Space Technology to the World Weather Watch, June 1967," *COSPAR Transactions* No. 3, 1967, COSPAR Secretariat, Paris. (Discusses the data requirements for global weather prediction experiments; techniques for remote sensing of the meteorological parameters with satellites; and data collection from atmospheric and surface platforms via a satellite.)

Computer Requirements:

Kolsky, H. G. "Computer Aspects of Meteorology," *IBM Journal of Research and Development,* Vol. 11, 1967, pp. 584–601. (A review of computer requirements, principally for numerical general circulation experiments.)

REMOTE SENSORS

Evans, W. E., E. J. Wiegman, W. Viezee, and M. G. H. Ligda. "Performance Specifications for a Meteorological Satellite Lidar," Note prepared for *Nat'l.*

Hero-4 Space Admin. by Stanford Research Institute, Menlo, Park, Calif., 1966.

Jones, R. F., *et al. Use of Ground-based Radar in Meteorology (Excluding Upper-wind Measurements,* WMO Technical Note No. 27, WMO No. 84, TP 35, 1959.

Katz, Y. H., Ed. *The Application of Passive Microwave Technology to Satellite Meteorology: A Symposium.* Prepared for Nat'l. Aeronautics and Space Admin. by the Rand Corporation, Santa Monica, Calif., 1963.

Kuettner, J. P., Ed. *Man's Geophysical Environment—Its Study From Space* Washington, D.C.: ESSA Publication, 1967.

WMO *Techniques D'Analyse et de Prévision des Champs de Vent et de Température à Haute Altitude,* Technical Note No. 35, WMO No. 106, TP 45, 1961.